Home Page of

www.sirnewtonsfruad.com

by Peet (P.S.J.) Schutte

ISBN-13: 978-1535240031

ISBN-10: 1535240032

All rights are reserved. No part, parts or the entirety of this book may be reproduced by publishing, electronically copied, duplicated by whatever means that form reproduction or duplication, without the prior written consent of the copy rite owner.

Written by Peet (P.S.J.) Schutte

This is the Home Page of

www.singularityrelevancy.com

http://www.lulu.com/content/e-book/wwwsingularityrelevancycom-website/8074920

ISBN 978-1-920430-58-0

© KOSMOLOGIESE EN ASTRONOMIESE TEGNIKA

All rights are reserved.
No part, parts or the entirety of this book may be reproduced by publishing, electronically copied, duplicated by whatever means that form reproduction or duplication, without the prior written consent of the copy rite owner.

This is the book showing everyone that there is A Conspiracy in Science in Progress

ISBN 978-1-920430-05-4 Written by P.S.J. (Peet) Schutte

© KOSMOLOGIESE EN ASTRONOMIESE TEGNIKA

mailto:info@singularityrelevancy.com

www.singularityrelavancy.com

In the web sight of
The Absolute Relevancy of Singularity
Which is Called
www.singularityrelavancy.com
shows that the cosmic phenomena perform the way they do by
Singularity

The Following Phenomena is Singularity forming a Universe

I don't only show what the four cosmic phenomena are being

1) The Lagrangian system 2) The Roche limit 3) The Titius Bode law 4) The Coanda affect.

This web site is not there just to show how the four phenomena are as they are because for that there are numerous other web sites specialising in that. There are many web sites that tells how these phenomena are as they are…I show why they are as they are and how it came to be that they are the way they are because as they are, they form the Universe as big as the Universe gets. To know why they are how they are is to know why the Universe is the way the Universe is and how the Universe came to be in the form it is. I don't only tell how they are as they are but explain conclusively each valid number filling the measure that they are where every number fills a place, and that is a first time in all of cosmology that this breakthrough is achieved. I don't merely mathematically formulate, but explain every digit as singularity forms the number and that is how singularity forms the cosmos. Every digit holds a number that validates a space filled that forms the Universe.

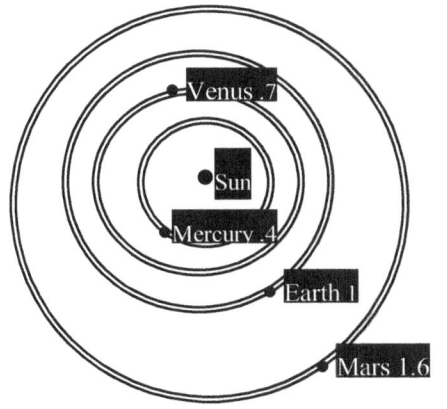
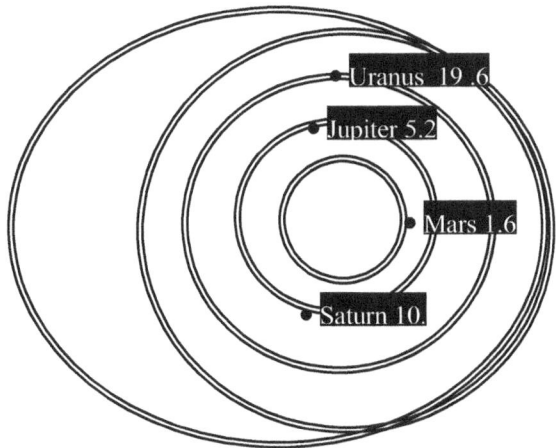

The Titius Bode Law in table form:

Planet	Mercury	Venus	Earth	Mars	Ceres	Jupiter	Saturn	Uranus
Bode's Law distance	4	7	10	16	28	52	100	196
Actual distance	3.9	7.2	10	15.2	28	52	95	192

The Titius Bode Law:

A numerical sequence announced by J.E. Bode in 1772, which matches the distances from the Sun of the six planets then known. It is also known as the Titius-Bode law, as it was first pointed out by the German mathematician Johann Daniel Titius (1729-96) in 1766. It is formed from the sequence 0,3,6,12,24,48,96, and 192 by adding 4 to each number. The planets were seen to fit this sequence quite well – as did Uranus, discovered in 1781. However, Neptune and Pluto do not conform to the 'law'. Bode's Law stimulated the search for a planet orbiting between Mars and Jupiter that led to the discovery of the first asteroids. It is often said that the law has no theoretical basis, but it does show how orbital resonance can lead to commensurability. The importance that becomes known is the sequence the Ties – Bode law saw in the number arrangement of 3; 6; 12; 24; 48; 96 etc. The incorrect application of the Titus Bode law lies in subtracting the figure of 3 from 10 leaving 7. The other way of reasoning is to add four each time to the firs value of three starting with 3 and so on. The true significance of the Titus-Bode law is that it points directly to a circular growth of 7 stages. The 7 relating to 10 is a precise derogative of the Roche limit or the Roche limit is a precise derogative of the Titius Bode principle because he two systems interlink.

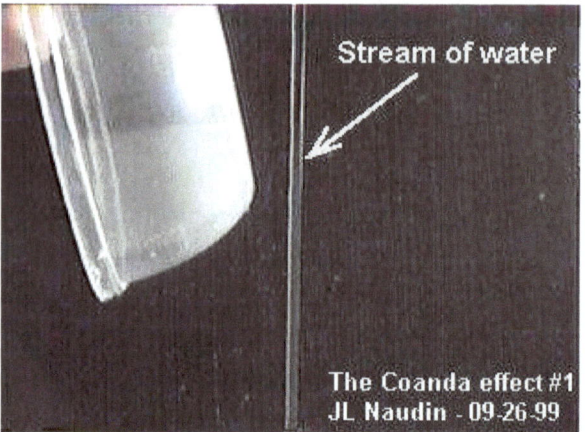 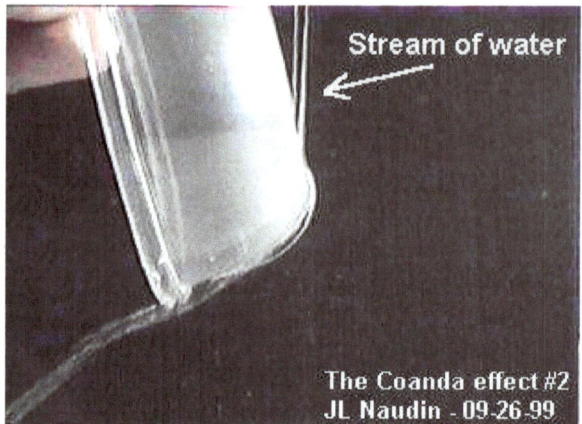

The Coanda effect

The Coanda effect applies as a gravitational phenomenon where moving liquid concentrates around the surface of round solid structures and by movement of either the liquid or the solid or both these concentrates the density of the liquid to gather and compact the flow of the liquid while remaining following the curve of the round surface. The liquid rather follows the curve of the round bowl than to fall straight to the Earth as on should expect. The liquid maintains relevance to the centre of such a round solid. I discard the idea that mass could be responsible for forming gravity because in almost four hundred years all evidence is indicating that the truth is to the contrary.

LAGRANGIAN POINT:

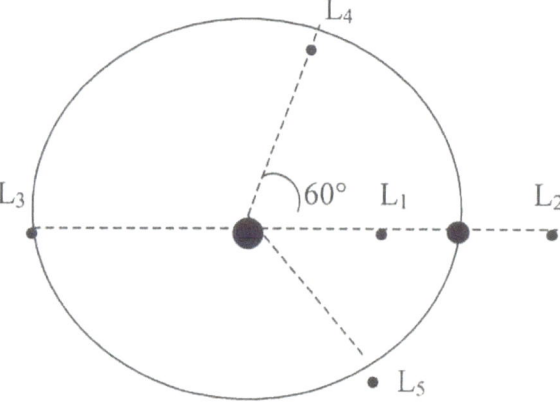

LAGRANGIAN POINT:
The Lagrangian points
are five equilibrium points
in the orbit of one body
around another, such
as a planet around the Sun

The phenomena are there and are applying! Put Newton's formula $F = G \dfrac{M_1 M_2}{r^2}$ to task and use it to explain these very common phenomena, and anyone would find it is not possible to use Newton and explain the gravity represented by this. The phenomena are there and applying so if Newton can't explain it then maybe Newton's concept of mass establishing gravity is not applying.

<u>It is this last statement where Newtonian science is unwavering in their believing that mass is forming gravity which is what I strongly bring into question.</u>

Please read on to find more information concerning <u>The Absolute Relevancy of Singularity</u>

The Roche limit is:
The region surrounding each star in a binary system, within which any material is gravitationally bound to that particular star. The boundary of the Roche lobes is an equipotential surface, and the lobes touch at the inner Lagrangian point, L_1, through which mass transfer may occur if one of the components expands to fill its lobe. It names after the French mathematician Edouard Albert Roche (1820-83).

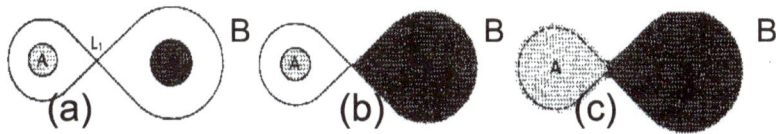

THE ROCHE LOBE: In a binary system, the Roche lobes of components A and B meet at the L_1 Lagrangian point. (a) In a detached system, neither star fills its Roche lobe. (b) In a semidetached system, one massive component, B, fills its Roche lobe. (c) In a contact binary, both components overfill their Roche lobes and share a common envelope.

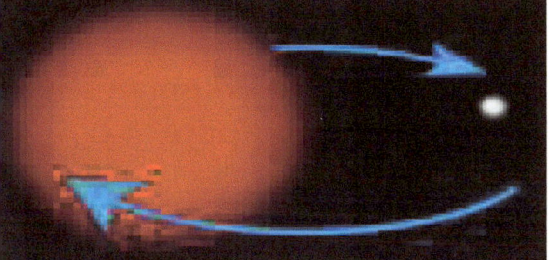  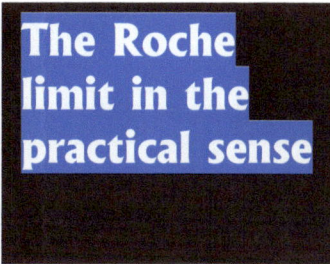

The Roche limit in the practical sense

HOME PAGE of THE ABSOLUTE RELEVANCY of SINGULARITY,
Forming the web site
www.singularityrelavancy.com

ISBN 978-0-9802725-3-6

My initial aim was to make this introduction article as simple as I could but I was forced to make it much more informing as I would and with it being an introduction article it now requires a lot more concentration to read than it should...

WRITTEN BY P. S. J. Schutte

All rights are reserved.
No part, parts or the entirety of this book may be reproduced by publishing, electronically copied, duplicated by whatever means that form reproduction or duplication of any description, without the prior written consent of the copy rite owner.

WRITTEN BY PEET SCHUTTE but belongs to
© KOSMOLOGIESE EN ASTRONOMIESE TEGNIKA

PART 2 of The Absolute Relevancy of Singularity,

Forming part of the web site **www.singularityrelavancy.com**

By going to LULU.com the following books are available in e-book format as individual books wherein I share with you the newly discovered information about

www.singularityrelavancy.com which you are reading and which you are free to download
Then download the next book from Lulu absolutely free and see if I exaggerate in any way!
The Absolute Relevancy of Singularity The (proposed) Article Free of Charge from Lulu
The Absolute Relevancy of Singularity The Dissertation
The Absolute Relevancy of Singularity in terms of Newton
The Absolute Relevancy of Singularity in terms of Cosmic Physics
The Absolute Relevancy of Singularity in terms of The Four Cosmic Phenomena
The Absolute Relevancy of Singularity in terms of The Sound Barrier
The Absolute Relevancy of Singularity in terms of The Cosmic Code
The Absolute Relevancy of Singularity in terms of Life

Should there be any person whishing to purchase these books in one volume given as a thesis published in paper format then contact me, on this web address by activating www.singularityrelavancy.com
And you will be able to purchase six books in print as a unit on paper forming one volume with six books going as *The Absolute Relevancy of Singularity* The Theses.
However please note that this printed Theses is very limited as it is printed privately. When you press on the button www.singularityrelavancy.com to activate I will return the e-mail as soon as I can to confirm the availability of published manuscripts and prices. The six books are identical to the six books on offer through Lulu.com but they are in monochrome whereas the individual book in e-book format are in colour where colour applies

The Absolute Relevancy of Singularity The Article is written as the first introduction to introduce singularity forming gravity in the new theorem explaining the Absolute Relevancy of Singularity. Since the article was comprehensive but was adjudged as to long for a physics journal, I decided to offer the article in its original and total layout in which I introduce the framework of my ideas.

The Absolute Relevancy of Singularity The Dissertation is there written as the second introduction to introduce the four pillars in a very wide sense on which the new theorem rests. This is to convince readers about the authenticity behind the explaining and the thinking that forms the new approach to physics backing the Absolute Relevancy of Singularity where gravity depends on Π.

Then The Absolute Relevancy of Singularity consists of a four individual part theses each forming a thesis. There are either six individual books on offer in e-book format or in print could only be purchased as one unit named The Absolute Relevancy of Singularity The Theses This consist of
www.singularityrelavancy.com which you are reading and which you are free to download
The Absolute Relevancy of Singularity The (proposed) Article Free of Charge from Lulu
The Absolute Relevancy of Singularity The (proposed) Article Free of Charge
The Absolute Relevancy of Singularity The Dissertation ISBN 978-0-9802725-8-1

The Absolute Relevancy of Singularity The Theses called
Thesis 1 The Absolute Relevancy of Singularity in terms of Cosmic Physics ISBN 978-0-9802725-2-9

The Absolute Relevancy of Singularity in Explaining the Sound Barrier called
Thesis 2 The Absolute Relevancy of Singularity in terms of The Sound Barrier ISBN 978-0-9802725-3-6

The Absolute Relevancy of Singularity explaining the Four Cosmic Phenomena called
Thesis 3 The Absolute Relevancy of Singularity in terms of The Four Cosmic Pillars
ISBN 978-0-9802725-5-0 and

The Absolute Relevancy of Singularity used to explain The Cosmic Code called
Thesis 4 The Absolute Relevancy of Singularity in terms of The Cosmic Code. ISBN 978-0-9802725-5-0

My mother tongue is Afrikaans, which is an African language and my second language is English which is the normal British /American/ Canadian / Australian variety used by many if not most. With English being my second language I am not boasting about my verbal skill in English and there is a hidden motive why I am mentioning that at this point, but I shall get to the explaining a little later on. Ever since my days as a student I had problems with accepting the logic behind Newton. In other words Newton's ideas about gravity never made much sense to me …and don't come with the nonsense that I don't understand Newton because after you have read my work you will have to admit I am the only one ever that understood Newton because I could correct his very flawed perception about gravity. Then after twenty-seven years of intensely studying gravity as a cosmic phenomenon I found the answer. I have followed a theory that I partly present in these books I named the Absolute Relevancy of Singularity The Theses, of which I investigated the research on a part time basis since 1977.

Ten years ago I decided to formulate my conclusions in a seven part theses I named Matter's Time In Space: The Theses Vol. 1 to 7 which I then compiled as my presentation of my new cosmic theory and then following that I worked on promoting my theses. This took almost every minute of my life the past ten years, as the promoting required my attention on full time basis whereby I was trying to introduce my findings to many academics without having much joy I should add. In promoting my ideas I wrote another twenty-seven books trying to make my ideas simpler or better understood. This past ten years saw me go without any income as I tried to get my theorem recognised, contacting institutions and intellectuals all over the world. Contacting people does not pay and you may take my word on that! Going without a steady income left me almost destitute and finally I decided to follow another path in order to find a manner to get my theory across where it will come to the attention of influential readers, and therefore I decided to publish these books electronically as to try and get around the stranglehold of Newtonian bias controlling science at present worldwide. I then decided to publish The Absolute Relevancy of Singularity The Theses as the six part manuscript by going electronically through LULU.com, which I saw as the only manner whereby I could generate funding by which I eventually would be able to have the twenty seven books I already wrote linguistically edited. Thereby I hope I will acquire the funding whereby I could have the books edited professionally and then afterwards have it published on a Print-On-Demand basis and then distributed through the large retail distributors such as Barns and Noble and Amazon.com. With my first language not being English and the books not linguistically checked by an expert there are bound to be language errors that readers will notice. In the past I tried to check my work myself but after checking say one hundred and fifty pages for language corrections, instead of having corrected work I ended by having four hundred pages of newly written information which is still not language corrected but holds a lot more information. This brought no solution but compiled the problem. This exacerbation of the problem is because my priorities lie elsewhere. I aim to spend money on correcting the work linguistically and then have the books formally printed in ink on paper, as I receive money and in the hope that I will receive money. I hope I will finally have all my work edited professionally as I hope I will find money to do so. However, the work I present I introduced for the very first time ever since time began and comes via my brain and every concept I offer, as an introduction to the world of science, is entirely a product of my mind. My promise to you is that if ever you are able to prove that the information I present as mine exclusively is not completely and altogether new, I shall personally refund your money immediately. I insist as I say as I prove that gravity forms by Π

www.singularityrelavancy.com which you are reading and which you are free to download; download it now

What you are about to read holds the dynamics that would change physics for all time to come. For the first time you will learn what gravity is as you would learn what singularity is as much as you will learn how to venture into a Universe that holds what there is together without mass, but all pulling goes by singularity arresting a bonded "flat" Universe in a state of gravity manipulating singularity. It is singularity holding the Universe together by moving time through space.

Gravity arrests the Universe, but no one ever managed to find the way it does arrest the Universe. If you read on you are about to find out how. I am taking the reader into a cosmos that holds a maximum value of 1 and anything greater than 1 does not fit into the Universe you are about to enter. Al the mind-boggling formulas used to impress has no meaning in singularity or in 1.

The Universe you are about to enter doesn't rely on a mathematical computing skills but an ability requiring human intellect through reasoning and following a line of debating. It requires the skill no computer could produce because it requires intelligent understanding of issues going beyond simply calculating and drawing unconsidered conclusions that is void of any intellectual understanding of cosmic principle, such as for example space whirls. If you have an ability to think and reason and don't require some mathematical disposition to rely on to help you think, then read on but be warned, this might be the highest intellectual level you ever called on.

Gravity that bonds the Universe together in the boundaries of singularity applying is a relation between material that moves in time and space that stands still and only moves by expanding. Awarding mass has no validity when objects form gravity but mass comes as a result of the above-mentioned relevancies.

Infinity is the centre line that can reduce no more and can never start, and that has no outside because it only has an inside

Eternity is the line that can't end because it can never stop expanding and therefore eternally moves and where it only has an outside because it can never reach the inside due to Π

To start explaining the Four Cosmic Phenomena we have to understand gravity. To understand gravity we have to understand the Four Cosmic Phenomena. Since science never yet came close to understanding the Four Cosmic Phenomena it is clear that science never understood gravity. Anything that spins forms a centre axis line I call infinity. Around this line a circle spins and the circle holds the value of Π and Π holds eternity away from infinity.

$$\Pi = \frac{21.991}{7}$$

$$\Pi = \frac{21.991}{7}$$

Gravity is the revaluation of Π in terms of the ring (7) and of the axis (Π°). In this the turning of a circle, the circle revalue 7 ÷ 7 = 1 *and* space revalue from 21.991 to 3.1416 = Π. That is how Π forms the curvature of space-time.

The axis that always forms a line holding 3 opposing points when the circle spins.

When a circle spins it forms an axis. Then the centre axis holds singularity @1. The rim of the circle is 7° where space then is 21.991. Gravity is Π moving from one dimension to another dimension and the Titius Bode law is absolute proof of this attachment that Π has to gravity. Gravity holds 7 in relation to 10. Gravity forms when 7 ÷ 7 = 1 and on the top 21.991 ÷ 7 = 3.1416. By compacting the space we establish a denser space we call the atmosphere and the atmosphere is the changing of 7° ÷ 7° = 1 and 21.991 ÷ 7° = Π. It is about movement of the Earth enforcing movement of space in a centrifugal pump action. In physics mass only pulls a cover over the eyes of those that are supposedly well informed intellectuals performing as qualified physicists that could in many centuries never, not once could explain how they say that mass has the ability to pull objects in the act of gravity.

The circle that always holds 4 opposing points when turning around the axis.

The Titius Bode law in conjunction with the Roche limit as well as the Lagrangian points conform to form a unit known as the Coanda effect. Science has no idea what establishes these four very important phenomena because science goes about studying the Universe incorrectly. No one could explain these crucial phenomena because the approach science uses to study the Universe completely incorrect.

The Titius Bode law shows the centre axis as a line that represents 3 singularity positions
The circle represents 4 singularity points or single dimensional positions.
The total in singularity is 7 and that is why Π holds in the circle 7 points and space as 21.991.
The space holds 10 points on either side of the circle (20/2 on the one side) in relation to the circle holding 7.
From the Titius Bode law one draws the proof that time holds singularity **a = (n + 4) / 10** and space is the result of time that moved on. One can see space is the footprints time left behind as time moved to the future leaving space as the past. That is the reason why the Titius Bode law, the Lagrangian points, the Roche limit and the Coanda effect forms the way entire the Universe unfolds and that excludes mass as a Universal factor altogether.

The sound barrier is not there to frustrate aeroplane pilots and warmongers. The sound barrier is the principle by which young stars start a life cycle and that is what is really important behind the sound barrier.

I explain the sound barrier according to the Titius Bode law, the Roche limit, the Lagrangian points and the Coanda effect and this is how the sound barrier unfolds. What happens in the case of the sound barrier is as follows: the Titius Bode law restarts the Universe as singularity arrests the Universe by applying gravity every instant time alternates.

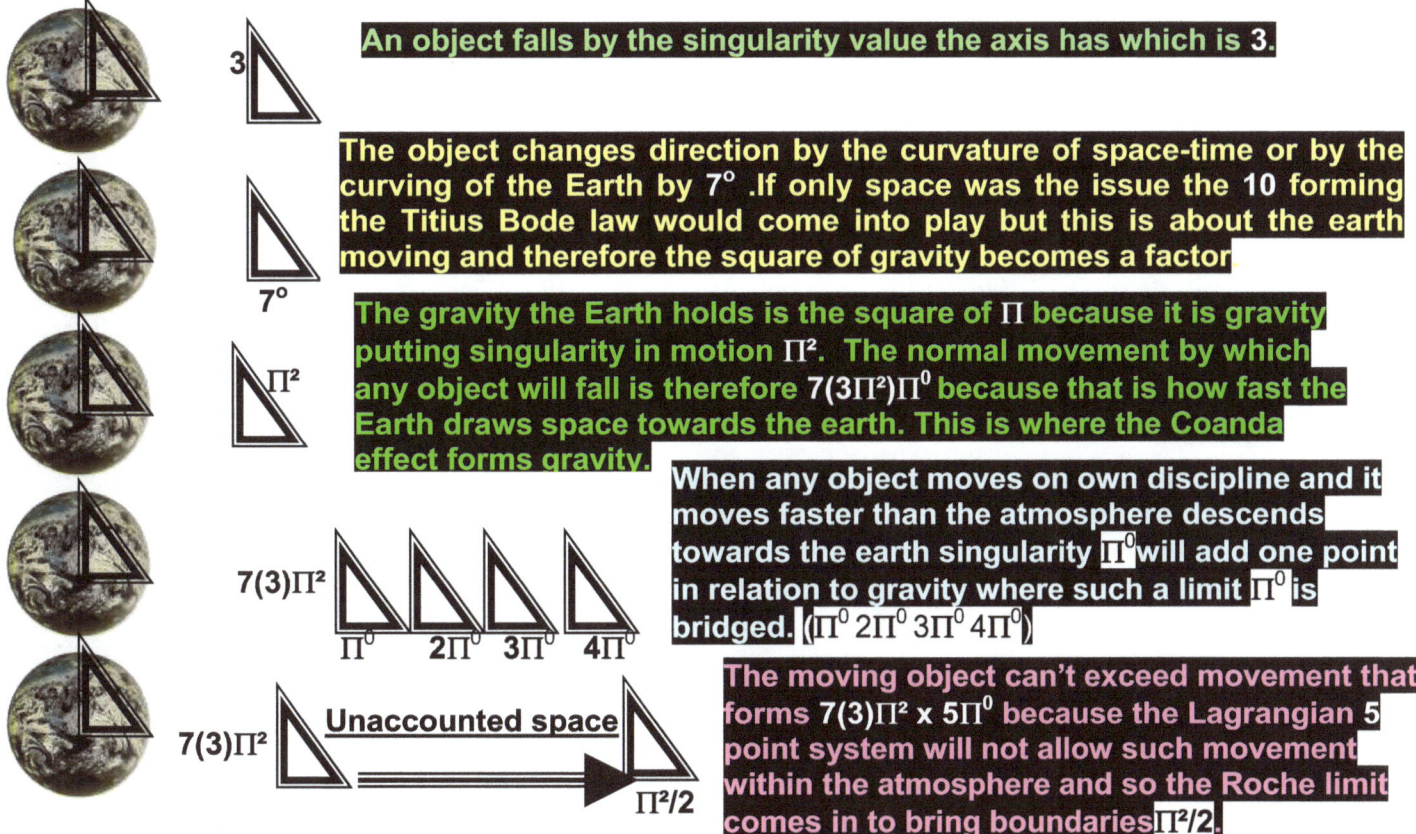

The Roche limit applying to stars being apart and the closest stars could be to each other is half singularity by the square which Π^2 is divided by **4**, which is, $\Pi^2/4$ but within the earth atmosphere this value halves to $\Pi^2/2$. So the sound barrier is **7(3)Π^2 x $\Pi^2/2$ =** 1022.79 km / h. To explain what happens when this happens and how this happens, the following gravity process applies. Gravity is the arrest singularity makes on the Universe in the single dimension where the highest value is 1.

Every instant singularity releases the Universe to space by restarting the Universe this goes according to the Titius Bode law bringing about space. When restarting takes place after the gravity arrest, then every point will be in place according to normal movement such as **7(3)Π^2**. The gravitational process places space in relation to material by movement of either or both. That is the Coanda effect. The higher an object is above the curving of the earth at the point where $\Pi = \frac{21.991}{7}$ changes to $\Pi=3.1416$ it has to be Π^0 away and the faster it moves (Π^0 to $4\Pi^0$) the higher the object must be from the earth curve. It is about movement away from the earth to maintain height in terms of motion just as a satellite must do and if not the satellite falls. The gravity shifts the points in singularity that forms space on position according to $\Pi = \frac{21.991}{7}$ going $\Pi=3.1416 +\Pi°$.

This puts space one point away in terms of singularity $\Pi°$. As the object moves in relation to the earth but not with the earth the movement must adjust the distance it moves according to the height it is in and this is **7(3)Π^2** from Π^0 to as much as $4\Pi^0$. The lines that $4\Pi^0$ forms are radial moving by increasing of 7° inclining towards the centre of the Earth. At a height of $4\Pi^0$ it will need to move is **7(3)Π^2 $4\Pi^0$** = 829 km / h to maintain space-time progress. At a height of **$4\Pi^0$** the speed of sound cannot be broken any more and this has nothing to do with sound. This is the transfer of lines carrying as much as connecting singularity to the centre of the earth. Then when the object moves faster than the allocated height requires movement should be a gap appears between the positions where the object should be and where the object then is much farther away. If the object moves as if it is at a height requiring **7(3)Π^2 x $\Pi^2/2$** an unfilled space appears that carries no sound because it is void of unfilled lines holding singularity at that moment.

Titius Bode law is formed from the sequence 0,3,6,12,24,48,96, and 192 by adding 4 to each number. The planets were seen to fit this sequence quite well – as did Uranus, discovered in 1781. However, Neptune and Pluto do not conform to the 'law'. Bode's Law stimulated the search for a planet orbiting between Mars and Jupiter that led to the discovery of the first asteroids. It is often said that the law has no theoretical basis, but it does show how orbital resonance can lead to commensurability. The importance that becomes known is the sequence the Ties – Bode law saw in the number arrangement of 3; 6; 12; 24; 48; 96 etc.

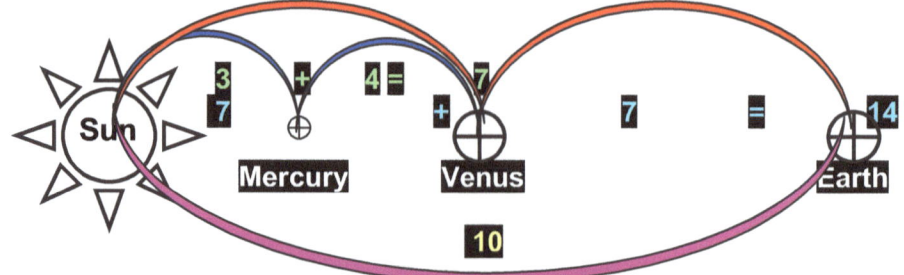

The space holding 7 plus the space holding 7 equals a position of 14. The circle is 4 as explained and the position of 14 minus the circle of 4 is 10. To place the earth according to a position in singularity in accordance to us on the Earth we have 10 of the earth divided by 10 in space giving us an allocated position of 1 This puts the Titius bode law in position with gravity and that proves that gravity is the forming of Π

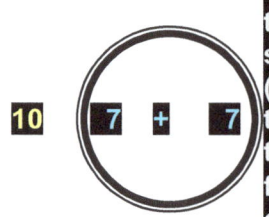

The curve of the earth is 7° on both sides (7° + 7°) but because 7° represents the earth turning in movement it is also ($7^2 + 7^2$) and by turning it crosses singularity (1^2) both sides of the opposing circle in rotation then according to the law of Pythagoras it is ($7^2 + 1^2$) + ($7^2 + 1^2$) = (49 + 1) + (49 + 1) = 50 on the triangle that forms by a circle turning the direction = 50 + 50 + 100. Therefore the space in which the circle turns is $100^{½}$ to the root thereof = 10 and therefore the Titius Bode law shows the inside of the circle factors forming Π as gravity. That is why 7 goes doubled minus the second part of the circle which is 4 divided by the space in which the planet orbits and the allocated singularity position according to the sun is derived. It is implementing Π as gravity

This is gravity. It is the way singularity freezes the entirety together into forming one, 1 atom, 1 earth, 1 solar system 1 Milky way, 1 Universe. The way gravity works is it puts the entire Universe in 1 relation for one instant and in that instant gravity applying is $7(3)Π^2$, the atom is $(Π^2+Π^2)(Π^2Π)3 = 1836$, where the Universe begins is the atom is $7/10(Π^6)+6=112$ and where the atom begins the Universe at the atomic start is $Π(Π^2+Π^2+Π^2+Π+3) = 112$. It is relevancies frozen in one moment by gravity where time forces the Universe to become an instant. The Universe freezes in the moment, like a movie film and then moves on to freeze again and that leaves nothing to calculate but to read and to interpret mathematical formula applying of relevancies applying.

So you intellectuals forming the accepting of standards in physics do not have a clue what I am talking about…that is because there are too much detailed explaining missing, the detail you never wish to read because words can't explain physics. For ten years no one wanted to read my work because I don't follow your guidelines when I explain things you have no idea how to understand and I don't use mathematics in formula to calculate the way you think physics should be…and yet you come no closer in understanding anything when using your calculations, notwithstanding your brilliant mathematical abilities and your overburdening egocentric arrogance but instead you lot criticize me for explaining the way I explain things you can't even think to explain, nor form any concept that could eventually lead to forming any conclusion thereof. You put yourself on a pedestal being bright and shiny insisting on the admiration of entire world and for me not following your principles you criticize me while I am able to show you things and give you knowledge that you can't buy with 3 billion dollars of tax payers money because your methods goes begging and still you find fault in my work because I don't compromise and follow your misleading methods. With your methods and the way you go about practising physics in the way you do you are getting no where for a long time and yet you will not read my work written in my way of conducting physics because it does not conform to the style you lot in physics wish to portray physics to be. If you will not read my work and accept there is another way of doing physics, then you will remain uninformed and feel proud about your ill understanding and your progress in stupidity. I enter singularity and you have no idea what it is except to give the concept of singularity as you see singularity to be a magical status. It never dawned on you that singularity is 1 in form and that is because you lot are besotted on your ability to mathematically calculate that which you can't even fathom.

PART 3 of The Absolute Relevancy of Singularity,

Forming part of the web site **www.singularityrelavancy.com**

From the Heart of the Author, the following:

Tyco Brahe, on his dying bed, kept on rambling to Johannes Kepler, begging Kepler to finish his work so that his life would not be in vain. Kepler complied but after five hundred years the work of Kepler is still unfinished and if I die without anyone reading my work, it is not only my work that will go unread but also again the work of Kepler as well, will remain unfinished, as it will never be understood in the way the cosmos intended. The Universe introduces a cosmos to Kepler that uses a formula that does not comply with the standards we see in the Universe that we see. Tyco Brahe spent a life time accumulating facts and the arrogance of semi blind, self-opinionated, self serving academics in physics of the day that thought they had the authority to decide what science is and what science has to be and what confirms science as much as what conforms science, decided Tyco Brae's studies were pointless.

Up to today the work of Tyco Brahe as well as Kepler is still unfinished and when I try to finish Kepler's work I find that my life's worth of work will be in vain, as modern mainstream physics will see to it that my views will never gets published because I question Newton. I do support Kepler but Newton never supported Kepler and because I support Kepler and not Newton, I am stupid and I am mentally retarded, slightly weak in thinking and mostly out of touch with reality. But that was how everyone in science at the time also felt about Tyco Brahe, Kepler and Galileo. I feel the same as Tyco Brahe, as Galileo Galilee felt and more, because I am bullied in the same way. Newtonians of the day keep going on about the Church and the wrongdoings committed to Galileo Galilee, but it is modern science that killed Galileo Galilee, because everything Galileo Galilee said Newton turned around and destroyed.

Gravity is no force…gravity is time because that is why one may employ the pendulum arm invented by Galileo Galilee to measure time. Time to science is how long it takes for the earth to circle once around the Sun, but for God sake, how can that time apply to the entire Universe. The pendulum swing you would use in the atmosphere of a massive star to measure the gravity applying will show time very different in the star. The thing about Newtonians is that they are not accountable to no one about what in their world contradicts reality. The one polishes the other and the lot shine while everything underneath is rusted to the core. And there I have lost the audience of every academic in physics, but I guess I never had their pompous attention anyway…

PLANET	PERIOD (Years) (T)	MOVEMENT (T^2)	DISTANCE	SPACE (a^3)	RATIO k
Mercury	0.241	0.058	0.39	0.059	0.983
Venus	0.615	0.378	0.728	0.381	0.992
Earth	1.000	1.000	1.000	1.000	1.000
Mars	1.881	3.54	1.524	3.54	1.000
Jupiter	11.86	140.66	5.20	140.6	1.000
Saturn	29.46	867.9	9.54	868.25	0.999
Uranus	84.008	7069	19.19	7067	1.000
Neptune	164.8	27159	30.07	27189	0.999
Pluto	248.4	61703	39.46	61443	1.004

KEPLER'S LAW OF PERIODS FOR THE SOLAR SYSTEM			
PLANET	SEMIMAJOR AXIS $a\,(10^{10}m)$	PERIOD T (y)	T^2/a^3 $(10^{-34}\,y^2/m^3)$
Mercury	5.79	0.241	k^{-1} = 2.99
Venus	10.8	0.615	k^{-1} = 3.00
Earth	15.0	1.00	k^{-1} = 2.96
Mars	22.8	1.88	k^{-1} = 2.98
Jupiter	77.8	11.9	k^{-1} = 3.01
Saturn	143	29.5	k^{-1} = 2.98
Uranus	287	84.0	k^{-1} = 2.98
Neptune	450	165	k^{-1} = 2.99
Pluto	590	248	k^{-1} = 2.99

In the tables that Kepler configured as $a^3=T^2k$ we have three distinct factors combining to form a specific value that indicates space-time $a^3=T^2k$ and moreover shows that the Universe structurally is composed in terms of **space a^3 = time T^2k** and every factor as much as a^3 and T^2 as well as **k** has a part and a role in forming the eventual value of **space - time $a^3=T^2k$**. The pendulum arm semi rotates T^2 in the gravity **k** of the space of the atmosphere a^3. For years science missed the principle that the pendulum measures space flowing in time thus **space - time $a^3=T^2k$**.
The part that mainstream science missed for five hundred years is the presentation of how the Universe is as it gave Kepler the formula of a "flat" Universe because the Universe we have is not substantial. Now all readers left

reading suddenly find this sudden urge to switch sides and begin agreeing with the Newtonians that I am somewhat soft between the ears. There the cosmos shows is a Universe in singularity that science only speculates about.

To incorporate a three-dimensional Universe into a flat Universe in singularity is not that simple. Whenever I am presented with this picture presentation of a flat Universe I am baffled by the Newtonian stupidity represented by 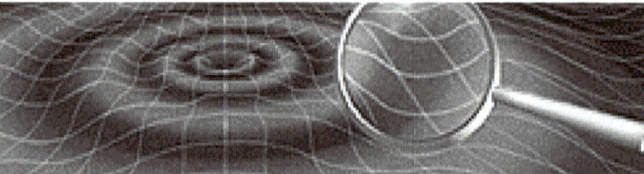 such an image. If there is a top, as the picture indicates as to be the mathematical flat Universe, then a bottom has to be somewhere, so it then is not flat. If there are waves, then there is depth and where there is depth there can't be flatness. Where there are blocks, then there is a width and a length and that can't be flatness because the blocks then show variation in size. This picture is as three dimensional as the blanket on any bed.

My question is if I see the topside, then mathematically there has to be a bottom, a side where the top ends because with the top ending in the picture I see there has to be a bottom side that is there but is only dimensionally obscured by the topside I do see. Secondly, the next question is what do I see? If it is the Universe the picture presents, then from where am I looking at the Universe I see and from what angle can I see the Universe in such a way because if I am, then I am part of the Universe in which I am. In this picture I am not in the Universe but I am a spectator of the Universe as if I am God looking down on creation. The only view I can have of a Universe is being part of the Universe and what they show makes me abstract from the Universe. I am always part of the Universe so I can only look at a Universe from up, down, back, front and both sideways forming a picture. I see waves. If there are waves, then there is a height and a depth bringing in dimensions complimenting the flatness they portray and that makes it a cube I am looking at. That then all this picture does, is it shows the funny way they see the Universe from the perspective they think they have using brilliant mathematics, as seen from what they think is their view of privilege by not being part of the Universe. This is the joke you will get when applying mathematics to a part of the Universe that holds all space as 1 because all the space that can fit into singularity is 1. Where singularity is one there 1 is the only applying value. If you don't recognise the fact, you get funnies such as they show in this picture.

I use the same top that Newton used when he tried to show that gravity forms by mass. I show with the top, an everyday object child's toy where to discover singularity. I show with the top where the Universe starts. I show with the top where it is that the Universe ends. I even show how to travel to reach the very centre spot in the Universe!

It's been five hundred years more or less and my study about the work of Kepler is the first authentic study ever conducted and because it is accurate and cuts Newton flagrant anomalies out, my studies are so disregarded I am told it is not science. In the studies of Tyco Brahe and Johannes Kepler they did not tell the Universe what they must find, should the Universe wish to be considered as correct. They studied the Universe and found the Universe told them that the formula formulating the Universe is $\underline{a^3 = T^2k}$. Never did they tell the Universe what the Universe should be, and that $\underline{a^3 = T^2k}$ is incorrect and it must be $\underline{a^3 = 4/3\pi r^3}$ and that is how one would measure a sphere, but instead learned from what the Universe told them and reported accurately. The Universe said it is… $\underline{\mathbf{a^3 = T^2k}}$

Does anyone see in the tables the Universe telling Kepler $a^3=T^2$. Newton told the Universe what the Universe had to be to comply with and what Newton wanted the Universe to be, it had to employ mass and when I ask to prove how mass is pulling the Universe closer, because even with the invention of Dark matter it still is not coming closer by mass… I am told I am not conducting science because science is conducted in a very specific manner where the Universe is told what to do. The Universe has to start shrinking because physics and science says Newton said it has to shrink…and now I too must comply or my work would never find publishing. Read on and find out how I am told what science is before science is physics. Science as Newton did, is still telling the Universe what to be…

Just the same way as the Universe are told it has to contract or be reinvented by dark matter changing the critical density that withholds the over all mythical contracting, I am told in the same manner what science is and how science must be conducted or my work will never find publishing. In science the protocol is the one academic is so much admiring the work of the other academic so that the other academic would admire his or her work rite back so that that they don't criticize the work of any body but criticize only the persons not echoing their views.

If you read on, you will see where the Universe starts and why the Universe can never end.
You will see precisely how the Universe goes "flat" because I vividly take you there.
You will see precisely why the Universe goes "flat" because looking at what I show you can see it with your eyes. You can actually cast your eyes on the spot that is while it also never can be part of the Universe.
You will see precisely when does the Universe go "flat" because you are part of the Universe going "flat".

And most of all, by giving you the picture I challenge you to show one point where I am incorrect or incoherent because what I say, everybody can read, if the person can read, because I am presenting you someone that apparently can't read, or so the academic claims because he says science can't be explained by using words…!

PART 4 of The Absolute Relevancy of Singularity,

Dear Dr. Schutte,

You submitted an article of 15 pages to the Annalen. The content of this paper doesn't constitute a theory in physics. With a lot of words and some simple algebraic relations, there is no way to "explain" the world of physics. You seem to be out of touch with modern developments. This is also shown by the fact that you don't quote any relevant literature. I am sorry to say, but the Annalen is not able to publish your work. I am sorry for having no better news for your.
Best regards,
Friedrich Hehl
Co-Editor Annalen der Physik (Berlin)
--Friedrich W. Hehl, Inst. Theor. Physics
* University of Cologne, 50923 Koeln _____/_____ Germany
fon +49-221-470-4200 or -4306, fax -5159
hehl@thp.uni-koeln.de, http://www.thp.uni-koeln.de/gravitation
* Univ. of Missouri, Dept. Phys. & Astr., Columbia, MO, USA
Dear Prof Friedrich W. Hehl,
I have received your e-mail reply and I wish to respond on your letter. This letter is a duplication of a previous letter but I repeat the letter because I selected material from my books to show you how much any one can prove facts and concepts in physics by applying "a lot of words and some simple algebraic relations, there is no way to "explain" the world of physics." In five articles I show you how you stray by using your mathematical calculations.

What I show you with these articles science with all the splendour of calculation and not using words could not explain in almost four centuries and that you will see if you care to read my work this time. I show you how ridiculous the use of the idea of applying mass are as Newton intended it applies in the cosmos. In sharp contrast I show exactly how the Titius Bode law applies gravity…and may I remind you that it is the Titius Bode law that is in place in the cosmos with the cosmos showing no evidence of mass playing any role in the formation of the solar system. Even in the arrangement of planets there are no evidence of mass favouring any position of any planet. However, truth for once be told, it is the Titius Bode law that undoubtedly forms the space that forms the solar system and I explain how it does it. I achieve this breakthrough that was never yet achieved by not trying to be clever, but by being honest about physics and searching within the cosmos for reality to bring facts to mind. I challenge you to show one word I use to explain physics and I add as it was never explained before, that does not become pure physics, basic physics as you have never experienced before, notwithstanding all your calculation ability. Everything concerning gravity goes by circles and in four hundred years all your calculations missed that.

However, if my explaining is not intellectually matching your understanding of the concepts I explain how gravity is Π and how the Titius Bode law forms Π and by doing that it also form gravity, where then this forms the solar system by the forming gravity, then do not merely reject my work unconditionally without reading it just because it is in your opinion The content of this paper doesn't constitute a theory in physics If you feel the content reduces your standard and belittles your ability to show any insight into physics, give it to a school child to read and let the child explain to you how the Titius Bode law forms gravity in the manner I prove that the Titius Bode law forms gravity by Π. Let the child also show where I prove it is impossible that mass can form a pulling power force as gravity because if it did, then by now mass would have had to destroy the entire solar system which it does not do.

The entire solar system is expanding and even the circumference of the earth is growing bigger, this concept ignores the basics of Newtonian physics. You can't support the Big Bang concept and at the same time hail Newton's mass forming a pulling force idea. It is not possible for the Universe to come from singularity at a point in the cosmos while at the same time mass contracts all other mass to a single point and if you believe that then reconsider your position in physics. The planets go by rings around the sun and not by mass pulling straight into the sun as Newton said mass pulls in a straight line. The only value that a circle or a ring can have is Π and that gravity shows it has. This information that you are about to read has more importance about cosmic physics than anything that you ever read before in your entire career. It shows for the first time how gravity evolves by showing how a phenomena forms gravity. I challenge you to prove only one idea wrong that I put foreword in all the articles. You are about to learn physics without the Newtonian bending of physics laws and see physics for what it is.

I am sending you seven articles wherein I prove that the validity of mass used in terms of Newton was never proven and wherein I prove how the Titius Bode law forms gravity. But true to Newtonian nature I presume all this information will again not be good enough and true to Newtonian character it will again be ignored because it will again not meet your Newtonian standards. If you do ignore my work again, it will be a blemish on your insight!

The article of 15 pages to the Annalen had in mind to introduce a very wide-ranging concept contained in many books. I wish to promote books in which I introduce a much larger and much more detailed cosmic picture. It is four books that actually form four volumes of one theme supporting The New Cosmic Theory. I wish to unveil a totally new approach to the thinking in cosmology. The concept is proposed in the article I sent to you which is "revealing" The New Cosmic Theory. In the article as much as the theme I wish to go where no one ever attempted to go before. I introduce the Universe of singularity, a state in which the Universe still is because it is a state from which the Universe grows. It is where material in a dimensional dynamic does not apply because it is where Einstein said "the Universe goes "flat"". I show you how and where the Universe goes "flat" I will guide you to the point where I go…so that you may see where my books and the article lead you. It is in the domain of singularity.

When you read work about the Big Bang you have to go right down the development (in reverse order) to the point where the Theory of the Big Bang points at a spot named singularity. It shows the very start from where all material developed. At that point one will find The Absolute Relevancy of Singularity and there has never been any attempt by any person ever to venture beyond the dimensional birth of the cosmos, which is called the Big Bang by going into the era where singularity prevailed. I take you there in my books as well as the unpublished article. However, going there requires a very high degree of concentration and calls for understanding that a very little number of persons are capable to show. I try to show how the Universe goes "flat" as Einstein said the Universe goes "flat". Even by completing this unimpressive letter you will also know how the Universe goes "flat". Even where you failed to read the article I sent you, then by just reading this letter you will be able to find where singularity takes the Universe "flat". But it requires a mental capacity to understand because where I venture no one ever in the history of mankind reached into before. I do not speculate but even in the unpublished article I show with pictures and sketches as well as "some simple algebraic relations" where to go to where the Universe starts, but you failed to read that because you are opinionated as to what conditions should the Universe have before the Universe will allow any one into physics. That is a pity. One should learn from the cosmos and not tell the cosmos what it must be to qualify as the cosmos. Then in the article I show you by almost taking your finger to the spot, the very point where the Universe ends and that too I qualify. You might dispute my arguments and show me about what you disagree, but it shows very little understanding of reason on your part about qualities man should have before understanding the Universe. I go into a Universe that was in place before light was in place in the Universe and only darkness prevailed because light calls for space and in that era of singularity space was not even a thought yet. I show why the Universe goes "flat" and in a "flat" Universe only the value of 1 holds value since singularity is 1. If you can understand 1 or $5^0 \times 7^0 \times 3^0 = 1^1$ you have all the mathematical skills required to understand the applying concepts. To reach a value of 1 does not require big mathematical equations but to reach singularity requires 1.

The collection I named The Absolute Relevancy of Singularity: The Theses and the collection as such forms a small introduction to the thirty-two or so books I wrote on various matters concerning physics with gravity in mind, but **The Theses** as such in the entirety of the four books does not officially even start to introduce the spectrum of every aspect of my work. I have been in contact with numerous Academics and about one in one hundred reply. When the one in a hundred reply, the academic always uses a most aggressive tone which I came to accept as what I receive from academics, and because of that I was most delighted to find some kind remarks from you as a practicing academic, and might I add, the first such kind remark in ten years of my trying to contact any person in physics that would take note of what I have to say about a new line of thought, because the few others that replied were extremely aggressive about me confronting Newton. I only began to submit books to publishers after twenty-seven years of studying Newton and the role Newton play in cosmology and thereafter which was ten years ago I began promoting these ideas. The New Cosmic Theory is a process wherein I try to introduce a study that is ongoing for about thirty-seven years, give or take a few and I did not jump into the frying pan having my first thought about the matter published as an article when I sent the article to the address of Annalen der Physics.

The New Cosmic Theory that I try to convey by writing books in total holds much information and every time when publishers reject the publishing of any entire book I propose, the rejection was on the grounds that "the discourse is not falling within the main-stream science discourse" and therefore I was subsequently advised to write articles on the subject as to find recognition. I was told that only then could I achieve publication of any entire book. Now I find that trying to publish articles has my work rejected on grounds as follows and the following is directly coming from the reply in which one of my articles was rejected recently: "You submitted an article of 15 pages to the Annalen. The content of this paper doesn't constitute a theory in physics. With a lot of words and some simple algebraic relations, there is no way to "explain" the world of physics. You seem to be out of touch with modern developments. This is also shown by the fact that you don't quote any relevant literature." It is not possible to introduce the totality of my work in 15 pages (or whatever a journal would allow) while remaining absolutely coherent on all aspects during such an introduction about anything. You wish for me to work with mathematics and calculations while the world I enter starts mathematics. My aim with the web site www.singularityrelavancy.com is to introduce the reader to a world before mathematics as a multiplying process took centre stage. I take the reader into the cosmic era when 1x1 was 1 and only 1+1 was valid forming 2. In the article I say that in so many words, and you would have noticed me saying this if only you took notice to read the article with care. I take you into a true flat Universe where space has no dimensions because dimensions are the multiplication of numbers whereas a flat Universe is found within the adding of numbers. Multiplying brings about a discipline of dimensions and singularity is void of dimensions, thus deemed to be single in dimension. The era we enter uses a line called time to create a single ongoing dimension.

I show why the triangle and the straight line and the half circle are all equal to 180° and in the world using space as form by using dimensions this fact about mathematics is bizarre. The triangle and the straight line and the half circle are all unequal in form while mathematics proves the three equal. It is obvious that the triangle and the straight line and the half circle are as wide apart as the sea and the Sun is, and yet there was a period in cosmic development when the three were mathematically equal as much as they still are. I have mathematics telling me this fact beyond doubt. Please use a formula and your brilliance in mathematics and using no words to prove to me why the triangle and the straight line and the half circle are all equal as they all are 180° while explaining details because on this rests one entire pillar of mathematics. The answer about this we find in the Lagrangian point system, which is one of the four cosmic phenomena, I explain when using the four cosmic phenomena to explain gravity. This becomes clear when using the law of Pythagoras to prove how this very law became the basis for mathematics and I do use mathematics in the law of Pythagoras to prove how mathematics started when the Universe started mathematics. However, I don't prove that in the article because the space allowed in the article is much to little to prove anything.

In the article however, I show why did Π become $21.991 \div 7$ or then $\Pi=3.1416$ or why is a circle Πr^2 or why is a circle circumference Πr or $\Pi d \div 2$. I show why a circle begins with Π and don't just surmise it. In my books I show why the phenomenon called the Titius Bode law is responsible for Π as a cosmic form and value. In my books I explain just as I claim in the article how the Roche limit come about and how the Roche limit is responsible for the sound barrier and what is the true cosmic value of the Roche limit as it plays a part in gravity on stars…that I show when I enter the era of singularity when calculations were still not yet developed. I show why a sphere in calculating the volume of space is represented as the formula $a^3=4\Pi r^3/3$ and why it is used to calculate the sphere when using these specific interpretations and how this is different from Kepler's $a^3 = kT^2$, which is the way to calculate volumetric space in applying singularity. The basis of this formula is derived from singularity finding form and that too I prove, but I have to use words because prior to when volumetric space came about, singularity prevailed and singularity is single dimensional. I pertinently state this over and over in the article. In the article alone I have no space to show all these facts and therefore in the article I only show why a circle uses Π to begin with. I show where and why did gravity start and what the true value of gravity is as gravity kick-started the Universe into a beginning because the beginning began with gravity. That I don't show in the article because printing space available will not permit me the opportunity to do so, but I introduce a book where I show exactly why, how and by which factors did the Universe start by using singularity. I show how the Universe evolved by singularity before space developed and at that time it implemented the four cosmic phenomena that later became part of space when space developed.

I am trying to introduce a study I have done during twenty-seven years of research and there is not one word that I can quote from any other source since every word comes from conclusions that I make and which I prove with the use of logic. All I try to do is to find a medium wherein I can tell some interested parties where to go to read my work and then for them to judge me on their merit and not be sidelined by rules set by academics in charge of publishing. Why don't you allow everyone to read my work and then afterwards, let all readers be opinionated by personal impressions applying and do evaluation of facts according to personal interpretations? Everyone goes on about the unfairness Galileo endured at the hands of the Catholic Church, but at least the Church allowed Galileo to publish his work so that the entire world could take note. Every one in science as well as the Church thought Galileo was out of touch when he declared the science wisdom prevailing at the time was incorrect, and five hundred years later we know who was out of touch as you state I am. I do not compare my work with that of Galileo but I find the same restrictions brought on me by the Powers of the day controlling science. The method of the blocking of getting new principles published is the same as what was in place back then where those in power controlled the thinking about science and those in power today still controls the thinking in terms of science by using equal draconian methods. By disallowing any other views to be printed that does not resonate with the prevailing mindset, science ensures that the public out there consider the correctness of their position as beyond suspect. Their discourse is then thought of as the only possible thinking policy that could be correct, which makes what they think absolute, beyond any suspicion any person could ever have. Killing criticism makes science deemed by everyone in the world as being undisputable because no one ever could dispute Newton. But it seems that no one ever got the opportunity to dispute Newton. Newton is only undisputable because disputing Newton is not permitted by science. Newton was never proven to be incorrect because any attempt to disprove Newton is killed in the infant stage and more often so even before birth of any such a thought could take place. I know this because for the past ten years the academic world holding publishing power destroyed every attempt I made to draw attention to the obvious insufficient work they base physics on. If you kill the messenger, no one will know about a new message and that is what happened then and that is what happens today. The Catholic Church was the one stopping Galileo, but nowhere is it mentioned that this was also in total collaboration with every party in physics at the time. Galileo did not only cross swords with the Catholic Church but crossed with the views the academics at the time had so Galileo went against what the academic world believed. It was the academics that prompted the Catholic Church to believe the Sun was circling the Earth and that the Earth was the centre of the Universe. Again I say I will not dare to compare my work with that of Galileo, but the treatment I receive I do compare. One thing science can take even less than the Catholic Church could is criticising their supremacy.

I have done twenty-seven years of research about the working of cosmology and found a manner by which I could interpret the four cosmic phenomena science do not even recognise because while they are there, they also don't fit into Newton's mathematical physics. As science goes, they will rather reject the obvious presence of the phenomena because it does not match Newton and must therefore be out of touch with modern developments. I did not only unravel the phenomena but worked out gravity from the manner the phenomena influence cosmology. The phenomena holds root in singularity and no one has yet entered that domain. All I try to do is to find a medium wherein I can tell some interested parties where to go to read my work and then for them to judge me on their merit and not be sidelined by rules set by academics in charge of publishing as Galileo endured. Let everyone read my work and then after that let all readers be opinionated by personal convictions applying. Allow my work to be evaluated by those reading it and not be smothered by those trying to kill the content because they do not care for the style I use. Galileo had an opinion that was clashing with the present dogma of the day but he could express his views because we now know about it. The way modern science kills me is they make very sure no one will ever know about me because they silence me as if I am dead. It is also so evident that at Galileo's trial academics were brilliantly absent by not showing a united effort to defend the liberty of thinking. That image today's science try to portray as if they now in all righteousness are fighting to uphold honesty. However, today one may only think freely as long as your thinking is echoing mainstream ideas. For ten years my ideas were constrained at every possible level I encountered and my ideas were destroyed, as Giordano Bruno was burnt alive. Before finding publishing I have to find favour in the eyes of the Academics in physics whom will not have my work published since I disagree with prevailing sentiment and I denounce Newton in terms of cosmology, but only in terms of cosmology.

That is what everyone misses.

Newton does not work in cosmology but Newton works in physics because in cosmology mass does not apply. In physics mass applies. I can find no evidence of mass doing anything in cosmology, still everyone grants mass because with mass it is easy to play with mathematics.

On earth where mass applies in everyday practice, Newton's work is undisputable correct but going into cosmology there is no evidence of mass applying, and that is where cosmology parts from physics. Mass do not pull planets and stars and that Hubble proved when Hubble proved the Universe is constantly expanding. I return to this elsewhere. Because I challenge everyone to show that mass plays a role in cosmology and in forty years no one could, my through thinking that my discourse is not falling within the main-stream science discourse and those with the power to prevent my work getting published will think up any excuse not to publish. They will block me because what I think will have modern thoughts prevailing in science at present brought into question. For forty years I have been asking that just for once someone will step forward to prove mathematically and without doubt that mass brings about gravity. Show the evidence that all the small stars are either in the centre of galactica or are on the outside of galactica and the arrangement of allocating stars go according to mass. What is it in the atom or the moon that has the ability to pull by magic something it does not connect to. Prove how it is possible that things fall by the measure of mass. Just for once show how things fall by the attraction of mass when everything proves that all things fall equal and therefore mass has no role to play in falling. The example used is a feather and a hammer falling in vacuum and this is fraud. Show how a car and a brick fall equal in front of a camera held by a cameraman and then tell persons the objects fall by mass issuing gravity proportionally according to the mass dishing out the gravity when the camera can follow both objects falling. If the ratio of mass brought about the ability of gravity pulling, then more massive things will fall faster and they don't. Mass does not pull or attract by any means or measure and also in this statement I return to debate it further elsewhere.

To bring one point to your attention just the following: you do support Newton and I question Newton and that questioning Newton is mainly what science hates. Where you underwrite Newton's claims of mass bringing the pulling of objects then please show me by using $F = G \frac{M_1 M_2}{r^2}$ how much did the mass of the earth draw the earth closer to the sun by using the mass of the Sun since the days of Kepler? You know as well as I do it does not happen because in fact the distance between the planets and the sun increases and does not decrease, as it should do according to Newton. Please use the formula that forms the basis of physics to show the world when will the BIG Collision come that will inevitably have to come if $F = G \frac{M_1 M_2}{r^2}$ is correct and when will the moon slam into the Earth. If Newton applies we await the collision between the earth and the moon because the masses on both ends will do the pulling to create the devastation that will follow the collision. Since Kepler made his calculations centuries ago, tell me how much did the moon come closer to the earth, presuming that $F = G \frac{M_1 M_2}{r^2}$ is indisputable. Did any member of physics ever bother to do such calculations as to determine when the collision is due, or have no one ever took interest in the case, and if not, then please tell why not. If the formula you mathematically base physics on was anywhere near correct applying in cosmology as you in physics claim it is, then you must be able to apply the formula and show the precise date such a collision will take place because every factor in the formula is known to science! Please show me on what evidence do you build your belief that mass pulls by other mass because from where I stand what I see is that science had to invent a graviton spawned from the imagination of science to try and address the question as to how does mass pull mass. I put it to you that your use of $F = G \frac{M_1 M_2}{r^2}$ is as correct as the presumption was in the time of Galileo that the Sun is circling around the Earth. Sir, the substance of power controlling thinking still prevails in science as much as it did in the days of Copernicus during his life where everyone had to submit to the thinking of the Powers in Charge of science albeit members of the Church back then for fear for your life as Copernicus did. One still do not dare ask questions or ask for proof as I do on the merit of mass as a factor in cosmology (not in physics) for then one would be silenced till death interrupts the questioning. Copernicus so feared not the Church but his colleges in science that he published his work after his death because then science could not kill him or employ the Church to do the official killing. Today science will allow me the privilege to die silently in a corner as long as I do it quietly because no one will allow my torturous screams to be heard. As in the time of the Copernicus, you lot still can't stand new thinking because new thinking will cultivate doubt in the minds of the many of those you consider as mindless and you require that they undoubtedly believe in you. By seeing to it that my work goes unpublished you willingly kill me by killing and destroying thirty years of my life…and then you lot point the finger at what was done back then as if you could be bothered by not implementing the very same evil. I ask you where is the freethinking and what happens to the freedom of speech as long as what is said is truthfully substantiated and can be proven because all the facts I present in the article you are unable to disprove and that is a challenge I put to you. Professor Hehl, you didn't even read my article because if you did you would see there is not one point in any argument about my work where I am incorrect, not one point you are able to disprove me or show me I don't follow the laws applying to physics in detail, yet you have to audacity to denounce my entire article just because it does not fit the profile you envisage it should. You and all your colleges are more condemning than the Pope and the Church was because at least they gave Galileo a fair hearing and considered his evidence. Even that you lot fail to do. I have had this treatment for ten years and every time it is the same over and over. You just couldn't be bothered to read it because it takes too much effort on your part to think in terms of evaluating every idea I put across, and there are a lot of new ideas you then have to chew on! What I touch on requires intellect to understand and not just some mathematical computing ability to perform. It shouts for human insight into cosmic forming that does away with make believe science such as Newton's guessing does with mass forming a factor and that the cosmos denounces. If you did bother to read the article, which you failed to do, you then would have seen I start where mathematics start and I can quote no one because I venture where no one has gone before; I go into singularity which by your definition is Singularity: a mathematical point at which certain physical quantities reach infinite values for example, according to the general relativity the curvature of space-time becomes infinite in a black hole. In the big bang theory the Universe was born from singularity in which the density and temperature of matter were infinite… and that I do quote, but that is all I can quote for the rest is the product of my labour and fruit of my mind. I challenge you to show where I stray from your definition of singularity when I show where to find singularity.

I explain just how it is possible to locate just such a point holding singularity to the precise value singularity must have in our modern Universe but I can assure you that where a mathematical point at which certain physical quantities reach infinite the grand splendour of mathematics are lost in dimensions not applying. I work in the era you can define but can't understand because it predates mathematics "singularity in which the density and temperature of matter were infinite" and it is in the infinite that mathematics becomes obsolete. Again I ask you that if no one ever has been there where I venture in physics, then whom must I quote because I quote the small part where science have been and that is all there is to quote. Every aspect about the Big Band deals with conditions prior to singularity deforming. If you use any quantity or formula based on numbers being more than 1 or any number to the power of zero, then you have left the realms of singularity because singularity could only be 1. At the point where singularity applies all complicated mathematics disappear. If you disagree, then give me any number that can apply to singularity other than 1 and please show me any mathematical formula that will apply to prove that singularity can be more than one.

If you did bother to read my article I sent you, you would have seen I show you exactly where $\Pi° = 1°$ could be found in the world of physics you study…and if I dare draw your attention to your accepted definition on singularity then as quoted it is a mathematical point at which certain physical quantities reach infinite" I show in the article where to locate this very point holding infinity. I also show where the point of singularity is infinite as it is holding what I named $\Pi° = 1°$. I show the point cannot ever start or become smaller since it is so small it has no space in which to form and if the point was in the Universe at the beginning, then it still has to be in the Universe because if it was in the Universe once, it must remain within the Universe because it has no other place to go by leaving the Universe. That is what you reject because that is what my article announces and my article introduces where to locate singularity and that is the article you reject. In this light going according to your attitude I am most delighted by your attitude, because from your attitude it is clear that where I venture you have never even left one thought.

In the web site www.singularityrelavancy.com I am introducing the reader to a world before mathematics as a multiplying process took centre stage. I take the reader into the cosmic era when 1x1 was 1 and only 1+1 was valid forming 2. This figuration proves mathematically that there was a time (1+1=2) pointing to a period before dimensions brought about perspectives (1x1=1). I take you into a true flat Universe where space has no dimensions because dimensions are the multiplication of numbers whereas a flat Universe is found within the adding of numbers and the adding of numbers point to a flat line forming the basis of the singular Universe. This process is directly formulated by translating Kepler's formula $a^3=kT^2$ to the true measure of gravity that is Π.

Please be so kind as to tell me, Professor Friedrich W. Hehl with all your mathematical splendour and magnificent abilities in constructing wisdom without using words, why is 1+1=2 and why is 2+2=4, because you do use it in physics, don't you? When you use numbers in your world of physics, being the Master that you are, you have thought about where numbers came from and how did numbers arrive? Please prove why it is that doubling two is also taking two into the square and while proving this, it is done without leaving a whisper of doubt. Please use your vast mathematical insight to explain without using words why would the third number be three, specifically three because that was how three came from singularity as 1 and why would the following number be four, which then is also the square of two by using the law of Pythagoras when proving this. How did five follow four to become five by using the law of Pythagoras to prove the point and then using this evidence to show that double five becomes ten, again by the merit of Pythagoras. Why would nine be the square of three and by adding 1 it becomes 10, because Professor Hehl, proving this is what really forms the basis of all science and that is how the Universe formed!

I wrote books about this process wherein I show numbers formed the start of the Universe and not material as you in science wish to believe. It formed by forming mathematics and the splendour of mathematics arrived only when the form of the Universe was completed and the cosmos stepped into the dimensional dynamics of space. This happened when the atom formed at $(\Pi^2+\Pi^2+\Pi^2+\Pi+3)=35.75 \times \Pi=112$ which is also when space as a whole formed at $7/10(\Pi^6) \div 6=112$. The relation $7/10(\Pi^6) \div 6$ validates the sphere as (Π^6) spinning in (7/10) a six sided cube ÷6 which is outer space. What I show when using $7/10(\Pi^6) \div 6$ was the moment the Universe came into dimensions by arriving at the formation of the atom. In the article I show why one might conclude why the Universe uses Π as a numerical basis for gravity, but I agree, the article alone does not start to prove anything because for that there are four books forming such proof called The Absolute Relevancy of Singularity: The Theses

I dare you, no I challenge you with all your mathematical splendour to prove one iota I produce as evidence in these books being incorrect, and you have to use words to prove me wrong because where I venture is where mathematics goes singular which was at a point before mathematics came in place…and that place can still to be located in the present Universe. I can and I do show you the very spot where the Universe came from, but not in the article for there is no room to do it. You, with all the astonishing mathematics are stuck at the point where the big bang arrived and at that point everything that forms the Universe was already formed within the Universe. The Universe adopted space at the event of the Big bang and therefore mathematical values came in a dimensional context at that point. However, everything that currently is, was already present in the Universe at the event called the Big Bang. Before that the Universe was one being 1^0 or 451^0 or $5^0 \times 1^0$ because that is what singularity implies the value of singular space is, it is 1. It is because you got stuck with your mathematics and you used your mathematics instead of brains and that is why that you can't proceed to resolve issues beyond where the cosmos formed the atom. You are all agreeing about everything coming from singularity but going there you have to lose your mathematical equations; it does not apply! If you in science realised mathematics construe singularity as one, science might have realised that mesmerising mathematics before the big bang was useless as a tool to formulate facts, then you would've realised how to reach a pre big bang Universe. I did just that and I can show how, and why and where the first moment arrived. I can show you precisely where that fist moment of arrival is today. The mathematics you apply had to start somewhere and it is there where I venture.

The Universe in singularity adopted the four phenomena which is called 1) The Lagrangian system 2) The Roche limit 3) The Titius Bode law 4) The Coanda affect but to unravel their meaning you have to go into singularity and to do that you have to understand Kepler and to understand Kepler that introduced singularity when he introduce $a^3 = kT^2$ forming the measure of singularity applying. You have to part what Newton thought Kepler said from what Kepler said and explain what Kepler really said, and that I do in the article by using some simple algebraic relations and by the meaning behind $a^3 = kT^2$ and those four phenomena the Universe came about. But the condition to understand how the Universe came about is to first understand the four phenomena interacting.

However, to understand how the Universe formed numerically or better titled **The New Cosmic Theory** one has to understand **The Cosmic Code** and learn how to read from it the interpretations of factors. To understand **The Cosmic Code** one has to understand the process of cosmic law supporting the Roche limit that works as what we think of in terms of forming **The Sound Barrier.**

To understand **The Sound Barrier** one has to understand the process of cosmic law supporting the Titius Bode law as well as the Lagrangian system that forms the Coanda effect and together the lot works as **The Four Cosmic Pillars** on which the entirety of everything was built by implementing singularity through a very specific process I named **The Four Cosmic Pillars**.

To understand those four comic laws applying one must be able to evaluate the process by reading **The Cosmic Code** in order to be able to recognise the actions brought about by singularity in relation to **Applying Physics** in terms of **The Absolute Relevancy of Singularity** and the one aspect of singularity applying in physics is being able to see how singularity forms space by the measure of Π, which is the only aspect of the entire collection of information I try to show in the article I sent to you…and you are unable to read that little bit…then how on earth will you ever get around to understand how the start of the Universe numerically came about when singularity and only singularity applied! Those books showing how the solar system was born and how the Universe came about I do not yet offer on sale. The information contained there are the really tough nutcrackers that explain by the cosmic code the inner working of gravity in stars and in galactica according to the cosmic code.

The Universe as we see it started much later in a period you call the Big bang but in truth there where the Big bang happened is when the atom came into form…and that I prove mathematically if you dare to read my work, which is the four books I wish to introduce via your Journal. Tell me Professor Friedrich W. Hehl, why is there mathematics formed by the adding process and mathematics formed by the multiplying process. In this evidence we find the development of the Universe. What happened that secured the forming of four to then be the prelude of five. Every number is a point and specific sets of points hold different relevancies placing the number in a quantification that brings about material in accordance to the coded relevancy. Please tell me the specific indisputable reason how the Universe did arrive at the value of what the number five depicts and it being precisely 5, or how did six become the next number on the numerical ladder and no other number but six tiny dots. Every number of dots serves a very specific purpose and five dots hold a value totally different from six. That is why nitrogen is a gas while carbon is a solid. That is why Mercury is a liquid with Xenon being a gas although they both are much more massive than say iron being a solid. Try to do this explaining according the law of Pythagoras by showing how the law of Pythagoras implemented the process and not use words. However, this is a country mile further than even the Cosmic Code is.

Why would seven bring about that a circle forms by redirecting directions as it is used in forming a circle by 7°? Why does the numerical value of 7° and only 7° play this role? Can you show me with your ingenious mathematics how the top part of Π is 21.991 when the bottom part is the circle by 7°, and use Pythagoras to substantiate the reasons. All the answers are in front of everyone but you said "With a lot of words and some simple algebraic relations, there is no way to "explain" the world of physics" and while you express you inability to see what I see, I see what I say I see and show what I see as clear as daylight while I do explain what I see precisely with a lot of words and some simple algebraic relations because complicated mathematical formulas did not yet enter the form of the Universe in the period I introduce as a The New Cosmic Theory. Before space applied, form applied and form was singular before space became dimensional. Have you ever considered why a triangle and a half circle are equal to a straight line by the dimensional value of 180° or is this the first time your intellect went that far? …And this question points towards your obvious mathematical brilliance in physics and not the lack thereof. This I point out to show while you do know everything about physics there is thought to be, there also is a small possibility that you do not know everything about physics that there might be. The Universe started numerically mathematical and not by material as material came later; everything used a numerical order to form.

In the article you failed to read I show precisely where the Universe goes flat and becomes singular with the little impressive "and some simple algebraic relations" I show precisely the point where the Universe goes flat and in line with your impressive mastering of mathematics I challenge you to use your mathematics to show where I am misinformed or where I fall from the wagon by using "and some simple algebraic relations". Guess what, the "simple algebraic relations" is what the Universe used mathematically to indicate to Johannes Kepler as to inform him as well as Tyco Brahe how the Universe is constructed. The Universe showed how the Universe used $a^3=kT^2$ which is some simple algebraic relations as a means of form, but true to your academic arrogance I see you know better than even the Universe does because the some simple algebraic relations $a^3=kT^2$ is what the Universe used to describe to Kepler about the form the Universe adopts. By using some simple algebraic relations $a^3=kT^2$ what is in the Universe became the form of the Universe, and you missed all of that…that is a pity. You know, using $a^3=kT^2$ I show where the Universe goes flat, a task no one ever mastered by using breathtaking mathematical equations.

Some academics previously indicated there was some point holding singularity within the black hole but I show in my books where this happens and where to locate singularity in everyday life. If Einstein said the Universe goes flat, then that flatness must still be around and be everywhere so that everything will be able to go flat every time Einstein said it does! However, the most wise amongst you failed to even value the measure of singularity, being everywhere and all around by using impressive mathematical equations let alone to pinpoint the point serving singularity and even less to indicate where the exact centre of the Universe are to be located. I show why the Universe is a sphere, which is a fact that is up to now only been surmised. I show why the Universe applies gravity as the form of the sphere, which is something all the brilliant masters in mathematics failed to deliver up to now. Why did all the mathematical masters fail to prove that the Universe uses the shape of a sphere while all pictures indicate everyone accepts the fact, thus failing to impress with your brilliant mathematics you have to surmise, as you have to do with most things. I dare you to read my article and show my arguments I present are failing in anything that I say it does. In the article I show where singularity applies and why singularity chooses Π as the form of gravity. Why singularity chooses Π as the form of gravity you are unable to prove when you are using those most impressive mathematical equations you refer to, because singularity does not apply impressive mathematical equations. Singularity applies simplicity.

You showed me that I "seem to be out of touch with modern developments". Please let me show you why I "seem to be out of touch with modern developments". Please let me show you what inconsistencies there is with the basic mathematical formula Newton introduced when Newton tried to use mathematics to prove that mass was responsible for gravity because you belittle the mathematics I apply **$a^3=kT^2$**, which is precisely the mathematics Kepler used to portrays how the Universe forms and that formula he read from the way the Universe is constructed. Then you look down your nose at Kepler's work while Newton's mathematics broke every possible mathematical law it can when Newton tried to convince the world how clever he was by cheating with mathematics.

Newton started off applying the factors holding in the relevancy as follows: $F = \dfrac{r^2}{M_1 M_2}$ and discovered it fell short of any form of accuracy. There is no way that this formula would ever work even by a lesser degree of accuracy.

Then Newton changed the formula to being the following $F \propto \dfrac{M_1 M_2}{r^2}$. Newton tried to convince (and succeeded) that one are able to change $F = \dfrac{r^2}{M_1 M_2}$ to $F \propto \dfrac{M_1 M_2}{r^2}$ while it meant the ratio would still work in the same way as if it was something like this: $F = \dfrac{M_1 M_2}{r^2}$ and the formula still didn't work. The changing of the formula you use as the corner stone, the foundation of all physics still proved to be a total disaster notwithstanding the cheating of the most fundamental mathematical law that should support all physics laws. Then Newton and his fellow boffins in science cheated mathematical law even further to change the lot to $F = G\dfrac{M_1 M_2}{r^2}$ without explaining how $F = \dfrac{r^2}{M_1 M_2}$ could end up as being equal to $F = G\dfrac{M_1 M_2}{r^2}$.

If you feel so strongly about mathematics used in physics then tell me Professor Friedrich W. Hehl, why don't you start to apply currencies to the factors and show the world how $F = \dfrac{r^2}{M_1 M_2}$ could become equal to $F \propto \dfrac{M_1 M_2}{r^2}$ and this equal ness could be carried on to eventually become the same principle as $F = \dfrac{M_1 M_2}{r^2}$ to then become $F = G\dfrac{M_1 M_2}{r^2}$. Put in real numerical values and show it does not constitute to mathematical fraud. If Newton were that correct, then please use the formula $F = G\dfrac{M_1 M_2}{r^2}$ to prove that the value derived from $F = \dfrac{r^2}{M_1 M_2}$ could eventually be the very same equal ness as one would achieve from $F = G\dfrac{M_1 M_2}{r^2}$.

Better still, why don't you write an article in your journal doing the song and dance about the accuracy the Universe proved Newton had when he implemented $F = G\dfrac{M_1 M_2}{r^2}$ because Hubble destroyed all the credibility that $F = G\dfrac{M_1 M_2}{r^2}$ once was thought to have. Then you can vindicate your attitude towards me while serving the cause of mathematics at the best you possibly can by restoring lost confidence in applying Newtonian religiosity. Put values to the factors and prove the Newtonian formulas have all the same results in the end. ...And while you are at it, write in the article showing how much did the distance there is between the earth and the moon reduce since the moon landing in 1969. Please use the most accurate figures available and then tell the world how much accuracy there is when implementing the calculation science applies to a shrinking Universe as Newton said it does when he said $F = \frac{r^2}{M_1 M_2}$ is equal to $F = G\frac{M_1 M_2}{r^2}$. Show how much all distances in the Universe shrink by their mass attracting other mass to bring about pulling forces of gravity applying throughout. Do use Newton's formula $F = G\frac{M_1 M_2}{r^2}$ to ensure accuracy. The Universe expands just as Hubble indicated it does as it expands from every centre holding singularity and it expands everywhere equally. Your colleague at Annalen Der Physics, Professor Doctor Ulrich Eckern once accused me of missing the basics of mathematics and classical mechanics by my evaluation of $F = G\frac{M_1 M_2}{r^2}$ but he failed to show what it is that I miss. Now it is the dream chance you have been waiting for...write an article about the correctness of the formula $F = G\frac{M_1 M_2}{r^2}$, show how the Universe comply to underwrite the absolute correctness of mass as a reliable factor shown by the formula having mass doing the pulling and then by the same token show what it is that I do not understand about the basics of mathematics and classical mechanics for I have been told this since my student days and after almost forty years I still fail to recognise what I am missing. To show me what it is I don't get, use the formula to prove how much did the moon come closer to the earth the past forty years by using $F = G\frac{M_1 M_2}{r^2}$ and the information acquired from data coming from the instruments placed on the moon for that sole purpose. If you are unable to do so, then never use mass in cosmology again because then mass is not pulling anything ever. Go one-step better... show why the Universe did not collapse back into singularity using $F = G\frac{M_1 M_2}{r^2}$ when r^2 was infinitely small and mass was absolutely contracted holding singularity in which the density and temperature of matter were infinite. Back when the Big Bang took place the entire Universe was in one Black Hole, then why did $F = G\frac{M_1 M_2}{r^2}$ allow the lot to escape. It will never again have that chance to bring everything that went loose back into contraction again. At that point the Universe had its best chance to collapse into the Big Crunch because the further the radius expands with Hubble expanding, the lesser the chance is that mass can do the pulling. With an ever-increasing radius by the square, the mass effectively will reduce in strength. If you prove why $F = G\frac{M_1 M_2}{r^2}$ did not pull everything back into singularity then that will be a worthwhile challenge for your brilliant mathematical skills to achieve! Then you lot can stop searching for the mythical dark matter that has to cover Newton's blatant errors because the dark matter is just a cover up. If the matter was there and is there presently it has mass, then why does it not use the mass it must have in the present to pull the Universe into contraction starting here and now and why is it waiting for something to unleash the forces of gravity by the mass of the dark matter. ...And by the way, why would the matter not pull now if it has mass just because it is dark... if it is going to pull it already has to pull or it will never pull because it is not there at all. It is much more likely to be in the imaginations and calculations of science that be in the actual Universe. The matter being in the Universe has to have mass, dark or not. If mass does the pulling, and it is there, it has to pull now, in the present at present or the entire idea is just another scientific hoax to cover Newton's incompetence. The matter being dark or not, has to have mass, visible or unseen and if the mass is there and mass pulls by the force of gravity, then please tell why the lot is not pulling now and what are the dark matter waiting for to start the pulling that will begin the contracting? I say this is more proof that there is no mass and that mass does not pull and the entire concept is to try and vindicate Newton's absolute misjudgement of gravity. It is one more compromise to cover-up science.

You see, Professor Friedrich W. Hehl, if mass was the factor initiating gravity or then the falling of a body to the ground, solid objects will have to fall faster than objects that is empty and hollow because the empty space within the hollow object will restrain the falling by not falling with the object since only the mass would tend to fall leaving the empty space behind to restrict the downwards descent of the falling object. If the emptiness within the cup did not fall with the cup falling then the emptiness will bring a drag on the solid part that falls. The empty part within the cup will try to stop the fall while a solid filled glass will then fall faster than an empty glass because the emptiness within the empty part of say the cup or glass falling would not fall, leaving only the small rim of the cup falling. With the major part not falling this hollow cup will fall slower while the fullness of any solid object will fall in its entirety, making the fall of the solid object unrestricted by having no empty space that does not fall and thus the solid object then will fall faster. Drop a full glass from an aeroplane with an empty glass as see the emptiness in the empty glass falls as fat as the water does in the glass. It is the space that moves down taking the body with.

A filled container does not fall faster than does an empty container and visa versa because the empty space of the object falls as fast as the filled space of the object and all objects fall equal and according to a variation in density in air caused by temperature fluctuation (excluding some gasses) allowing any variety of mass to fall equally. It is the space and all the space notwithstanding being filled or not that falls or moves towards the roundness of the earth proving that space holding material or not holding material falls equally notwithstanding mass and for that reason that is why Galileo's pendulum swings regardless of pendulum length or size as Galileo said it would. It is the descending space driving the pendulum that swings.

This again was proven by the very first ever experiment concluded scientifically. This fact of space descending does not come as a surprise because Empedocles proved this fact back in 450 BC. Empedocles showed that space displaces water from the clepsydra, which was a sphere shape container with a sprout on the top and small holes in the sphere through which water ran in small streams out at the bottom. When the flow of air or space was blocked in the spout by a finger covering the hole at the top of the sprout at the entry, the water stopped flowing from the clepsydra. They concluded in 450BC that it is the empty space that pushes the water out of the clepsydra because the moment one restricts the empty space or air to flow into the clepsydra from the top, the water will stop flowing out of the bottom of the clepsydra. Why would the flow of the water stop if the mass did pull the water down? When the finger blocks the sprout and stop the space entering from the top, the water does not fall to the ground but it is the empty space that pushes the water out at the bottom to fill the clepsydra from the top. When the finger blocks the sprout and stop air to come in through the sprout opening the water should still run out at the bottom by the mass of the water pulling, if mass was doing the pulling. If mass was the force giving factor, then the water must keep on flowing because the mass of the water did not disappear when the sprout was covered and therefore it still has to produce the pulling by forming gravity. All this evidence was known to science about 2500 years ago but since "With a lot of words and some simple algebraic relations, there is no way to "explain" the world of physics" it lacked mathematical communication and it should therefore surprise anyone very little that physics could not fathom this result 2500 years onwards. Professor Friedrich W. Hehl, do try and find the ability to use a lot of words to "explain" the world of physics because it is helpful preserving past experiments and results as it then does broaden one's horizons mentally…sometimes!

Forget the example always used about the hammer and the feather falling equal in a vacuum because the hammer and the nail and the elephant falling together will also fall equally notwithstanding falling in a vacuum or not falling in a vacuum. The vacuum part is conspicuously in place to purposely confuse reality as it is brought in to flagrantly spread misunderstanding of the issues in hand about the falling that takes place. With everything always falling equally when the same the condition applies to all objects falling and therefore with such falling happening under the very same variation of natural conditions applying, this shows it is the space in which the object is that falls and not the object falling while leaving the space it holds behind. The lack of relevant density in relation to air moving down stops the feather from falling equal just as gas does not fall with the space at the rate that space does descend. People realised this fact 2500 years ago but then used a lot of words to "explain" the world of physics and today because of not using a lot of words to "explain" the world of physics science has no idea how to interpret the very first experiment ever conducted! That is a travesty as much as it is a tragedy. All space falls by the compressing of the atmospheric space. The rotation of the earth moves the space sideways and this brings the space to move downwards by increasing the density of space or air as it comes closer to the earth. This results from the Roche limit applying to fix atmospheric layers varying in density. In my books I explain that principle applying mathematically. The increase in atmospheric density is the result of the rotation motion of the earth brought on by the Roche limit applying while it takes filled and unfilled space towards the solid of the ground and that is what the Coanda effect shows which is what your brilliant mathematics in one hundred years could not begin to explain. The Coanda effect is around for almost one hundred years and please use your mathematical skills to explain why the water will rather follow the roundness and flow with a detour along the rim of the glass than fall straight down as it should when mass would pull? That is the principle behind the sound barrier, and the Coanda effect, and wind restriction and hurricanes and more other things than I have room to mention. With all the attempts made in that past to uncover those issues I mention, it never was resolved notwithstanding all the impressive mathematics available to use. Notwithstanding using your mathematical marvels, science has not got any vague idea to explain any of the phenomena mentioned above. That is why modern science seems to be out of touch with modern developments. To understand these phenomena one has to understand singularity.

The simplicity singularity applies is shown when I show why the triangle and the straight line and the half circle are all equal to 180° but when considering form using mathematical dimensions this mathematical fact seems bizarre. It is obvious that the triangle and the straight line and the half circle are as wide apart as the sea and the Sun is, and yet there was a period in cosmic development when the three were mathematically equal as much as they still are.

In the books, not the article, I show by using the law of Pythagoras why did Π become 21.991÷7 or then is Π=3.1416. I show using the law of Pythagoras how and why by the law of Pythagoras is a circle Πr² or why by the law of Pythagoras is a circle circumference Πr or Πd÷2.

I show mathematically by the law of Pythagoras why is a circle using the specific value Π has to begin any circle or sphere with, but due to lack of space I can only prove it in my books as I can't prove it in the article. I show where and why did gravity start by the law of Pythagoras and what the true value of gravity is as gravity kick-started the Universe into a beginning. Can you show how the law of Pythagoras was implemented when the Universe formed, because I can show the reasons why and I do show the reasons why with using words since the law of Pythagoras implemented actual basic mathematics? I show why the law of Pythagoras implements the law it carries. The reason for this is the method how the Universe started off and the reasons why the Universe began. Maybe you should try to use words one day; after all it is a helpful tool in explaining physics, because it surely helped me explain what was never explained before. I mention the law of Pythagoras because the Universe does apply the law of Pythagoras in all of the cosmos and therefore the law of Pythagoras is part of physics, don't you agree? Can you use your breathtaking mathematics without using words to tell how did it come about that the law of Pythagoras has the dynamics it portrays it has in mathematics as well as physics, because I can by using words. Use your astonishing mathematics to show why everything started by the law of Pythagoras. Mathematics can't do it because the law of Pythagoras forms mathematics and the law of Pythagoras helped to form mathematics as mathematics developed. The four cosmic principles yet not understood by science shows just how Pythagoras applies.

Mass has nothing to do with falling and all things fall equal as if having equal mass when falling because all things fall equal in relation to the space in which they are. It is by buoyancy that space holds things and that removes mass while falling as a factor. Space not holding things fall with space holding things while it is therefore not the mass that causes the falling but the compressing of space which you call the atmosphere. The falling is written in relation with the value of Π. The value of Π is 3.1416÷1 or it is 21.991÷7. It is not coincidental that Π has two distinct equal values because it is due to precisely that that the first moment in the Universe came about when point 1 parted from point 2 putting space in-between eternity and infinity.

There are two values forming Π as much as confirming Π. The air or space holds 21.991 when the Earth holds 7° but when spinning the earth applies the change of direction by instating the axis by the value of Π°Π which is the centre line or axis or earth centre Π° connecting singularity to the earth circle Π. Then relevancies in Π changes as the space that was 21.991 with the air held a link to the roundness of the earth being 7° at the time. But as the 7° dived into 21.991, the 7° goes singular or 1 as the space then in turn becomes Π=3.1416 or becomes the circle. The earth and the circle of the Earth becomes (7÷7)=Π° or 1 while the rim of the earth is Π=3.1416. Forming Π=3.1416÷Π° the form Π then aligns with centre of the earth holding singularity Π°because the axis placed singularity Π° in centre stage when the earth turned. Every time the earth or sphere turns, it places the surrounding space in relevancy from 21.991÷7 to form Π°Π. All this I said in the article that you say is not physic s because I use of a lot of words to "explain" the world of physics By the way, now for the first time in you entire career you also know what gravity is and what forms gravity and doesn't it make a lot more sense than to presume mass pulls mass by gravity without having a stitch of true evidence to prove Newton correct?

What happens is that the space condenses (21.991÷7) by the turning of the planet or star and the compacting of space surrounding the spinning sphere results from the rotational movement (÷7) of the earth that brings about that this compresses the space of air or atmosphere (21.991) into more density (÷7) which is done by movement of the space surrounding the turning object (21.991÷7) albeit a planet or a star and moves space filled with whatever or unfilled, going vertically down towards the roundness of the Earth that then is represented by the circle or the rim of the Earth (Π=3.1416÷1). Everything within the concentrating space will come closer to the surface because it is the space that moves down to the earth and not only the object filling space, but everything within the space including the space that falls downward. Every micro millimetre the relevancy of space changes from (21.991÷7) to the roundness of (Π=3.1416÷1) until the Earth forms the final (Π=3.1416÷1) and the object finds mass as a result.

The falling body never stops falling but find that mass comes about when relevancies changes from Π=21.991÷7 to Π=3.1416÷Π°. By touching the Earth, and by that ending the relevancy of Π=21.991÷7 from reapplying, the object then becomes part of the earth circle Π=3.1416÷Π° and having contact with the axis Π°Π it becomes part of the Earth singularity distribution and only then finds in this relevancy applying the reward of mass. The body never stops falling but as the earth by density restrict the body movement vertically according to density, the falling becomes a tendency to move downwards in order to unite with singularity formed within the centre of the spinning earth. This is all about relevancies changing and relevancies reapplying positional changes, which is what gravity or time is. This is how the Universe goes flat or singular. It is Π°Π and that I say so many times in the article that you were incapable to read or you refused to read or you were not motivated to read…you can make your choice about you're reasons withholding you to understand what I explained in the unpublished article. Regrettably this is "a lot of words to "explain" the world of physics and because of the use of a lot of words to "explain" the world of physics it seems to be out of touch with modern developments making the process of thought very difficult to comply with modern developments and prevent many academics this far to understand and therefore you lot would rather cling to the use of $F = G \frac{M_1 M_2}{r^2}$, which is truly what seems to be out of touch with modern developments since everyone accepts that the Universe expands. I prove my point of gravity being Π using some simple algebraic relations…you now have the chance to prove me wrong by proving how correct the cosmos shows Newton and his mass is. Show that Jupiter is coming closer to the Earth and therefore we have to relocate to some other galactica or die! Show why Jupiter is randomly located notwithstanding size and why the planets do not arrange positional allocations by the implementing of mass as the factor that would and that should arrange the positions of the various planets should mass truly be a reality in cosmology.

The spin of the sphere constitute of a change in direction to the value of 7°. In the centre of the circle the axis are Π^0 and therefore the spin makes the directional change reform to singularity or change to the value of the circle in relation or in relativity with the axis in having the circle 3.1416 and having the centre or singularity or the axis $\div \Pi^0$ and with that connection the circle, which is space, goes singular or goes flat bringing about the much argued flat Universe. This is how gravity puts multi-dimensional space going into singularity or Π^0. This means that the 7° becomes one or singular and space changes from $(21.991 \div 7)$ to $\Pi\Pi^0$. On the top of the equation the value of Π is 21.991, and by revaluating 7 through spin to become 1 that value changes to the value of 3.1416 being in relation to $\Pi^0 = 1$. The rest of the explanation that will bring proof to my statement when using Pythagoras is far too bulky to offer it at this point. I did say it in the article you refuse to publish because you refuse to read it that the relevancy of gravity is the changing of the value of $\Pi\Pi^0$.

The space reduces $(21.991 \div 7)$ to conform to singularity $3.1416 \div \Pi^0$ by the rotation of the sphere that produces an axis by initiating singularity Π^0. In the books I show the very reason why is $\Pi = 21.991 \div 7$ and I use the law of Pythagoras underwriting the Titius Bode law, that conjuncts with the Roche limit as well as the Lagrangian points to prove the Coanda effect and the Coanda effect, as I show in the article you didn't read and therefore wouldn't publish, is gravity by principal. Gravity is the Coanda effect that is the changing of liquids $\Pi = 21.991 \div 7$ to form solids $\Pi = 3.1416 \div \Pi^0$. I much rather say you didn't read it as putting it down to you not understanding it because I use "a lot of words to "explain" the world of physics and which possibly tops your understanding limitations and therefore you didn't publish it, don't you think…? It will be of no use to explain how the law of Pythagoras was implemented to prove $\Pi = 21.991 \div 7$ because I use words which you so honestly admit you don't fathom and it takes far to much space to explain the entire process in this letter. However, I do show how I mathematically conclude this value by using the law of Pythagoras in the books, if you care to look.

With the gravity being Π and that gravity comes as a result of the earth's spin contracting the space forming a denser substance called the atmosphere, which comes about and in accordance to every sphere spinning around an axis of its own, the increase in density around every spinning object brings about a loss in the overall density of space between all cosmic objects such as the earth and the moon and all the planets and the sun and the loss in overall density brings along that the distance controlled by the density of the substance there is between all cosmic structures gains in space. That is what the Big Bang was from the start. It is the substance that fills space in-between cosmic objects is that what you mathematically see as dark matter and it is limiting absolute expanding at the rate of what the density will allow. It is working as it should at present and no search is required. With the Big Bang the density in unoccupied space decreases as the density in occupied space increases making the Universe to seemingly expand, which it can't do. It is relevancies reapplying. The earth can't pull the moon by mass.

Mass does not constitute to the falling of objects but the space compressing brings about the falling action notwithstanding the space being filled with objects or empty of objects and therefore as Newton so vividly proved, one can pull a cover over the eyes of all but one (and that one is me) by cheating with mathematical formulas but to cheat by using words are a lot more difficult. The only thing Newton pulled by mass is a huge cover over many eyes for a long time. All things fall equal as we see on TV everyday where people and cars and bicycles and beds fall equal when dropped from aeroplanes and therefore it has to be the space in which the objects are including all the surrounding space in which the objects are not that is falling and not the object because of mass or having more favourable density. …And don't say I don't understand or I miss Newton because if you say you believe $F = G \frac{M_1 M_2}{r^2}$, then surely if you might understand Newton because then you clearly seems to be out of touch with modern developments in terms of cosmology and applying physics. It was shown this past century that the Universe is expanding and not contracting by mass…expanding means the lot forming everything in the Universe are drifting apart but that is "a lot of words to "explain" the world of physics which explains why modern Newtonian science seems to be out of touch with modern developments.

But all persons filling academic posts in science holds the attitude that they know all and others know nothing and academics know best while the rest of the population is mindless, thoughtless and worthless. Professor Friedrich W. Hehl, you are no exception to this rule. In fact, you are one of the best examples I have seen. I have taken these insults long enough…its been raining on me constantly for ten years ever since I tried to introduce my first thesis and if you wish to insult me, do so while you see there are in science parts that is yet still undiscovered and rather try to find how much there is that you don't know about science instead of thinking how much you as a scientist knows about science and what great achievement science is instead of trying to go where science still has to go. Let me give you some wisdom. A wise man thinks of all the things he does not know and what awaits his discovery while it is a fool that thinks of how much he knows and feels impressed by his personal field of knowledge. The best answer to a question one can have is "why" since no answer ever brings full conclusions and science missed some.

Yet you lot in science seem to know you've concluded everything that anybody could ever uncover about science. Recently by some academic in physics blew me away when I tried to introduce him to The New Cosmic Theory. I am afraid that you will continue to get rejections if you do not relate your work to existing theories and previous work. While it is possible that a lay person hits on an insight that has been overlooked by academic trained in the field over many years, it is unlikely. We assume that work offering something new would be related to existing theories, either by building on top of them or by showing how and where they fall short. If you do not relate to existing work, it is repeatedly going to be dismissed as mind spin too easy to shoot down. I am sure you understand. This is what I have been told when my work was again rejected by another non-complying professor. When I show mass plays no part in cosmic physics by building on top of them or by showing how and where they fall short my methods don't apply. When I show how to enter singularity "a lot of words to "explain" the world of physics So you lot block anything constructive not coming from mainstream science because you lot are the only ones knowledgeable about science. I have shown how knowledgeable you lot are and what it is you protect by pushing me off the table. You say I don't understand Newton while I am the only one ever that understood Newton. I understand very well that Newton doesn't work because mass influences physics but plays no part in cosmology. Show me how does mass pull because I prove what gravity is and why gravity pulls by Π, but you lot refuse to read because that is an easy way to escape…that is also a cowardly way to protect what you try to hide! Tell me why did it take you lot since 1860 or thereabouts not to be able to explain The Roche limit because I can, but I have to detour from normal science. In modern science you confuse physics applying on earth with how the Universe works because Newton got the lot confused. The physics on earth uses mass but cosmology in the Universe applies singularity, precisely as Kepler introduced it by introducing $a^3 = T^2k$ which in singularity is $k^0 = a^3 \div (T^2k)$.

You lot are not even able to start to explain The Titius Bode law, a law that is in place since it was discovered in 1766. The law is so vital planets are discover by applying this law and you know nothing constructive why this law is in place or why it arranges planets as it does. You ignore the law because it ignores mass. By this law I can explain how gravity works. Mass has no part in cosmology as it applies in the Universe, but singularity rules the lot.

Has any one in your league have any clue why The Lagrangian system forms as it does while applying your most impressive mathematical equations. It's been around since 1772 and in almost three hundred years you have not even come close to any attempt to explain the phenomenon. Studying this system shows singularity rules the lot.

The Coanda affect is the way all jet propulsion works on and all aerodynamics and wind restriction works on, and yet you have not even come close even to find an inconclusive explanation, even in a feeble context of any sorts. Yet when I use thirty two years of my life to find a method to explain these phenomena that then bridges the barrier preventing science to enter the era before the Big Bang came about, there is only locked doors preventing me to find publishing. The centre of a solid "pulls" closer the liquid by Π and the earth is solid with the air being liquid.

What are you lot scared of? Are you scared of what you hide would be uncovered and that the people's admiration would tarnish when your dark methods are unveiled… for dark they are because on the most critical there could be in science, how mass pulls, that most basic you only assume? You lot in the modern era are many times over more protectionists than the Roman Catholic Church was at their worst. The Roman Catholic Church thought they represented God and acted as such. You lot think you are God and act accordingly and therefore you allow no one to know anything if you lot did not know it first. You are unable to tell me what the four phenomena are, let alone explain their function and purpose in detail. When I try to introduce a road that will lead to a method whereby one may uncover the purpose and the working of gravity and the manner in how the four phenomena forms gravity and how the four phenomena kick-started the Universe into what it is, everyone in science blocks me because I represent the devil since you are god. You stop me from showing anything new because then I show why you lot are wrong. You might hold all the power, but when using your power you only advance stupidity, as this recent project in the mountain of Switzerland will once again prove. You'll learn nothing because you are going about the wrong way. The Big Bang started with numbers and relevancies that then by applying those dots in relevancies became material and space. If you read my article that I sent you, you would have seen the detail how it happens.

I am no longer taking these insults on the cheek and riding it out as I did do so many, many times during the past ten years while trying to get my idea of The New Cosmic Theory read by anyone that is not so sublimely self-opinionated as I find practising and teaching academics in physics are. I try to introduce the cosmos, as it was in the pre-big bang era when only singularity prevailed because if it did prevail then it still has to prevail just as Einstein surmised it does in the flat Universe he saw. If it was part of the Universe back then when singularity prevailed, then it still must be part of the Universe since it has nowhere else to go when it leaves while it has nowhere to leave. The cosmos is written in a mathematical cosmic code and I found a way to translate the code and using that I became able to understand many unresolved facts about the cosmos. Please be warned that there is no simplicity such as just awarding mass in the forming of gravity.

The process is immensely more complicated than awarding mass because this is done without the circumvention of the truth by sidestepping reality in proving the factor that mass is forming gravity by a factor in alignment of cosmic objects having a specific volume in size or in any cosmic planet showing more pulling power. You can't just simply gauge an object and then award mass by the size it has to cheat reality, because in cosmology there is no such an escape root. In fact size plays no role in the cosmos because the smallest star there can ever be, the black hole, is also the biggest star there can ever be. If you disagree with my statement that mass does not apply in cosmic terms, then show what role mass plays in the positional allocation of planets circling the sun. See how planets align in the solar system and from their size, prove they use mass to line up accordance to mass being a factor... I prove that gravity is Π and every circle every planet makes is vivid proof of the fact that that gravity forms by Π applying. If the planets as well as the sun pulled by mass, then surely the lot had to be part of the sun by now because the pulling has been going on for some time. The truth is the planets are drifting further away from the sun every instant of time and how do you reconcile that with Newton's force of mass giving a pulling power.

The proof I bring is true about gravity being formed as a result of implementing the following phenomena, **The Lagrangian system** 2) The Roche limit 3) The Titius Bode law 4) The Coanda affect, and the combination these phenomena we find the sphere as a multi-dimensional circle present the form Π, which I explain by delivering mathematical proof as to how they fit into the overall picture of gravity. I prove the fact that every individual one of those phenomena is forming a unit that is in total being what we think of as gravity. The phenomena altogether constitutes a unit that forms the process working as gravity. The phenomena are there and with them applying in physics how can anyone dare as to be so arrogant in saying my work is not physics when I explain cosmic laws.

I am going to use this letter as a web page introduction in the future as to show anyone that wishes to read my work, where I stand on matters concerning science. I think it gives a splendid opportunity since this shows the story of my life the last ten years and my encounters with all academics in physics! That makes this letter an open letter.

From here on the web page starts

The Absolute Relevancy of Singularity: The Dissertation The book with which I hope to convince all the non-believers and general doubters about the authenticity of my new approach to science. Should any person feel a need to first find conviction in my claims I make, I have published this title to introduce my thoughts. Should the interested party need convincing about the authenticity and the new approach I take to science before committing to purchase the four books, then this book will do all the convincing required by any sceptic.

What you are going to read is new to all of mankind. Everything you are going to encounter was never written before. Whether you are the most accomplished physicist or a first year student fresh out of school you have the same background knowledge about the work you will encounter. It is advisable therefore to first get acquainted with the first, and then the second book then the third book before reading the fourth THE COSMIC CODE.

There is a Universe in difference between the top lying down and a top spinning erect and all of that difference

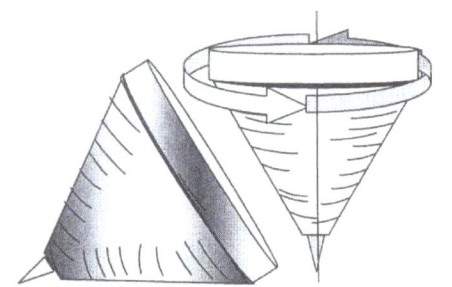

Newton missed three and a half centuries ago. In the books I explain in much detail (as well as all the others I have written on the matter) as well as that I prove that gravity forms through movement in relation to these four cosmic phenomena and the movement establishes Π which translates into gravity. By movement gravity applies and in that mass is a by-product resulting from gravity moving and then forming mass in the process. Mass is the result of gravity and is not the factor producing gravity. Mass results in gravity forming.

Gravity forms by the spinning movement of material and in that establishes singularity that is then initiating a circle that produces singularity by using Π. That is gravity.

By the movement of the top that is holding the top erect, the top in moving fights off the mass that the Earth bestow on the top and where mass kills movement by having the top lying down still on the Earth, the moving top comes erect as it then fights the Earth's gravity by applying its independent individual gravity.

By applying individual motion to the status of the top on top of the structural individuality that increases by the motion that the Earth provides, the independence of the individual object is becoming further exaggerated by having independent motion, which is further defying the incorporation the Earth strives to achieve. As the motion of the independent object grows more independent by applying more excessive motion to such an extent where motion creates almost the ultimate independence that may free the individual object with independence from the motion the Earth creates is what is breaking the restraint gravity has on all objects with independence formed by their structure. The structure show independence at all times by not forming part of the structure of the Earth within the sphere of the Earth's gravity. Moving about shows even more reluctance on the part of the top when spinning allows the top to eventually become part of the Earth. Breaking the sound barrier is the motion in space duplicating space by crossing over gravity borders, which is the limit to what constraint the Earth may produce in accordance with what full independence would allow.

The phenomena form an intergraded unit that results in gravity forming where each forms a part of gravity. You may still be you would be sceptical …but convince yourself that I did manage to:
1) Find the location, position of singularity as a factor forming space-time
2) Finding space-time by dissecting Kepler's formula in relation to valuing singularity
3) Finding and proving space-time and aligning space-time with gravity
4) Find the working principals behind gravity as a cosmic occurrence.
5) Find the reason for the Roche limit and explaining the resulting of gravity from that.
6) Find out why the Lagrangian system, becomes the building form of the Universe.
7) Find why the Titius Bode law mathematically provides the foundation of gravity
By proving that the Coanda affect is gravity through activating space-time
By using the above the four cosmic pillars, it enables me to present the proof where I now can explain what conditions bring on the sound barrier. By proving it is gravity that the individual structure generates motion above and beyond the gravity the Earth provide is what is producing individual motion that the independent object earned within the sphere of motion that the Earth's gravity provides where the independent and individual motion put the relevance that gravity has beyond the conserving means gravity has where the space that is serving the independent object is independently in motion. The adding to the independence on top of the normal structural independence is creating more individualism by the independent motion of the individual structure being apart from the motion that the gravity of the Earth provides. The fact every one misses is that any structure that is not part of the Earth's crust has an independent gravity and the form this gravity applies is stronger than the Earth's gravity which is why the structure maintains its form and this provides the independent individuality the structure has giving the unique structural space. The gravity of the Earth strives to incorporate everything into the Earth's sphere and into the Earth's structure and therefore the fact that the object is not incorporated into the Earth shows defiance and individuality, which gives it, mass.

These are the definitions underwriting cosmology and while my work is that much ignored; read my books and let's see how far I stray from these definitions in comparison of how much Mainstream science underwrites these definitions by them bringing indisputable proof in presenting unwavering hardcore facts as I do. The manner that Mainstream science interprets these definitions is a joke not worthy to mention.

Quoted directly from the Oxford dictionary of Astronomy the following:
The definition of space-time is as follows:
Space-time is a four dimensional position of the Universe where the position of an object is specified by three coordinates in space and one position in time. According to the theory of special relativity there is no absolute time, which can be measured independently of the observer, so events that are simultaneous as seen from one observer occur at different times when seen from a different place. Time must therefore be measured in a relative manner as are positions in three-dimensional Euclidean space, and this is achieved through the concept of space-time. The trajectory of an object in space-time is called world line. General relativity relates to curvature of space-time to the positions and motions of particles of matter.

The definition of singularity is as follows:
Singularity: a mathematical point at which certain physical quantities reach infinite values for example, according to the general relativity the curvature of space-time becomes infinite in a black hole. In the big bang theory the Universe was born from singularity in which the density and temperature of matter were infinite.

The Oxford dictionary of Astronomy defines gravitation as follows

Gravitation is the force of attraction that operates between all bodies. The size of the attraction depends on the masses of the bodies and the distance between them; gravitational force diminishes by the square of the distance apart according to the inverse square law. Gravitation is the weakest of the four fundamental forces in nature. I. Newton formulated the laws of gravitational attraction and showed that a body behaves as though all its mass were concentrated at its centre of gravity. Hence the gravitational force acts along a joining of the centres of gravity of the two masses. In the general theory of relativity gravitation is interpreted as the distortion of space. Gravitational forces are significant between large masses such as stars planets and satellites, and it is this force, which is responsible for holding together the major components of the Universe. However on the atomic scale the gravitational force is about 10^{40} times weaker than the force of electromagnetic attraction

In the books on offer through this web page and in which I am introducing a totally new concept in terms of gravity, the proof I bring is true about gravity being formed as a result of these phenomena. In the past science hardly recognised the existence of such phenomena although they are known to science for centuries.

They are known as

1) The Lagrangian system
2) The Roche limit
3) The Titius Bode law
4) The Coanda affect

However, since the explanations that I provide holds a completely new line of thought, there are just too many and too numerous wide ranging facts behind that which forms the complete picture as a whole, this leaves me unable to include a full introduction in a space as small as that which a web page will allow. The explaining of such a totally new approach includes for instance those phenomena science this far failed to understand and which I have named as the four cosmic pillars. With these facts being altogether new to science, I find academics showing very little willingness to consider the acceptable value thereof. I recon it must be the result of science seeing so many idle explanations in the past and then proving to be senseless as much as being little impressive, therefore my mentioning it without bringing and substantiating proof will be fruitless and counter productive.

I found the manner in which to interpret Kepler's formula as $a^3 = kT^2$ and I found that when dealing with Kepler's formula, we should not see a^3 as space but we should see singularity being positioned in space in relation to singularity forming relevancies. What brought the answers was putting singularity in context with Π. Doing that placed me in the position to discover what gravity is and how gravity operates to form the Universe. I saw that Kepler's formula should instead use Π. By placing Π in relation to gravity I manage to find an explanation for the four cosmic phenomena. Everything that has anything to do with gravity forms a circle albeit that it is called the curvature of space-time or gravity bending light or forming a round galactica, the connecting factor is gravity which implements Π. Gravity or another name used to call gravity would be time is running on the measure of Π and every aspect of cosmology integrates Π as the basic concept on which cosmology is founded.

Because my views do not echo the commendable praise attributed to the greatness by which Newton is commemorated, my work is purposely and very much wilfully poorly received in the world of physics and astrophysics and by that I find very little willingness in any understanding shown in the ranks of Newtonian science. This work contains ideas about the introducing of a totally new concept on explaining for the first time ever the working principles of gravity, a matter that eluded Newton no less. I decided to offer four books that introduce the explaining of these concepts in e-book format. This method of publishing rests totally on a financial basis. I tried to introduce the four phenomena as a concept by using a web page but found such introduction is far too comprehensive in having just too many and numerously wide ranging facts that form the complete picture as a whole to be comprehensibly appreciable, and therefore on account of that realisation that I was unable to include a full introduction in a space as small as that which a web page will allow, it gave me the idea of introducing this new concept via electronic publishing. As my other books I sell by printing are all hundreds to thousands of Mega Bites of information, I had to revise the layout where each is to have fewer than twenty mega bites. This motivated me to only introduce the concepts in producing small books that then could be sold via the electronic publishing media as to allow persons to first acquaint themselves about the viability of the concepts and the feasibility of this new approach I introduce. If any person shows interest in finding out more about any of the books, please click on the book of interest and discover something in science no one yet has ever heard about.

The main issue of finding the value and the meaning of the four phenomena was to connect gravity to Π. Gravity is much closer connected and is much more intimately related to Π than it ever can be linked with mass. By giving each of the phenomena a measured value in terms of Π solved every riddle connected to the phenomena and not only did the phenomena become purposely clear but also the working principle gravity…

What you purchase is information and not ink put on paper, forming a copy of yet another science book. You purchase information never yet divulged, and I am not exaggerating. Look at what new knowledge that I uncover as I explain for the first time the following phenomena:

The Roche limit is:

The region surrounding each star in a binary system, within which any material is gravitationally bound to that particular star. The boundary of the Roche lobes is an equipotential surface, and the lobes touch at the inner Lagrangian point, L_1, through which mass transfer may occur if one of the components expands to fill its lobe. It names after the French mathematician Edouard Albert Roche (1820-83).
\

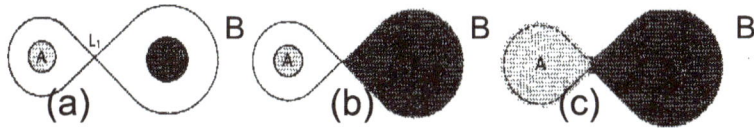

THE ROCHE LOBE: In a binary system, the Roche lobes of components A and B meet at the L_1 Lagrangian point. (a) In a detached system, neither star fills its Roche lobe. (b) In a semidetached system, one massive component, B, fills its Roche lobe. (c) In a contact binary, both components overfill their Roche lobes and share a common envelope.

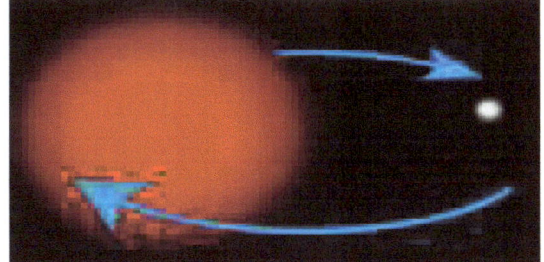 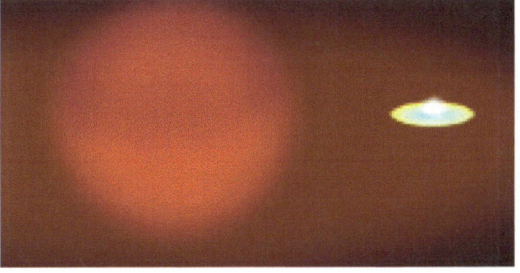 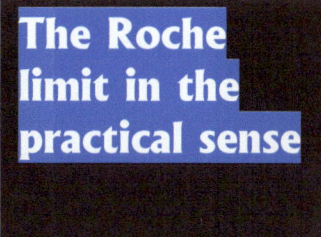

The Roche limit in the practical sense

The following will offend Newtonians beyond what words can describe but the truth is there for all to see. The formula $F = G\dfrac{M_1 M_2}{r^2}$ cannot explain the comic occurrence shown in the pictures above called the Roche limit, I should find some attention when I say I can explain what is occurring in this instance and this occurrence connects directly to the Roche limit, as explained above. Not only does the Roche limit explain this phenomenon, but also it ties directly to the Titius Bode principle, also being another inexplicable factor in light of the formula $F = G\dfrac{M_1 M_2}{r^2}$. The Roche limit liquefies the minor structure when the two are 2.4674 x the radius apart and in the process the major star absorbs the minor star after in ballooned the smaller one to the radius belonging to the larger one. This is kept under wrap because $F = G\dfrac{M_1 M_2}{r^2}$ can't explain what happens while it discredits Newton completely. I can explain it, but only after I abandon Newton's $F = G\dfrac{M_1 M_2}{r^2}$

According to the formula of $F = G\dfrac{M_1 M_2}{r^2}$ all orbiting structures should collide with a bang, but instead it is evident that they do the tango until one drops, but when dropping it still does not collide with the larger structure, instead it is liquefied and than is absorbed. There is no form of collision ever taking place as would the formula $F = G\dfrac{M_1 M_2}{r^2}$ suggest that is used by science.

The position where the formula applies is most surprising. Where the formula $F = G\dfrac{M_1 M_2}{r^2}$ applies, one has to find singularity applying because the position of r is pointing to a specific pinpointing of space contracting.

This is not only limited to planets in our solar system. In the Universe, there are giant stars spinning around each other. These stars are binaries, which are also one form of double stars where double stars are another such a form. The difference between the types depends on the distance they remain apart. They keep a certain distance apart and do not collide. In the case of the sun and its planets, it could be a case that the systems might be too small, or they might be too apart. However, this is not the case with binary stars. They are close, they are big, and they spin around a mean axes called the Roche limit.

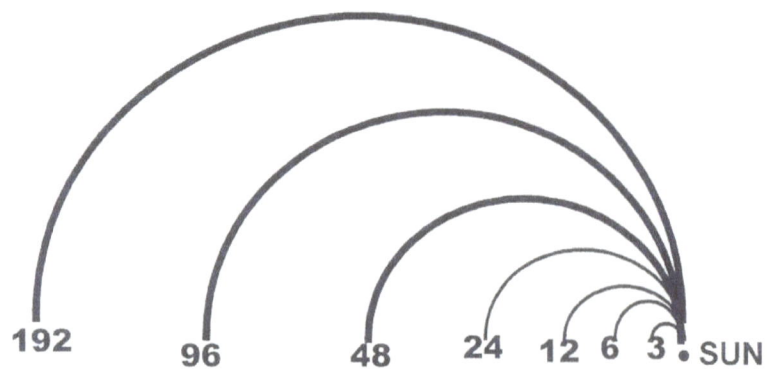

The Titius Bode Law in table form:

Planet	Mercury	Venus	Earth	Mars	Ceres	Jupiter	Saturn	Uranus
Bode's Law distance	4	7	10	16	28	52	100	196
Actual distance	3.9	7.2	10	15.2	28	52	95	192

The Titius Bode Law:

A numerical sequence announced by J.E. Bode in 1772, which matches the distances from the Sun of the six planets then known. It is also known as the Titius-Bode law, as it was first pointed out by the German mathematician Johann Daniel Titius (1729-96) in 1766. It is formed from the sequence 0,3,6,12,24,48,96, and 192 by adding 4 to each number. The planets were seen to fit this sequence quite well – as did Uranus, discovered in 1781. However, Neptune and Pluto do not conform to the 'law'. Bode's Law stimulated the search for a planet orbiting between Mars and Jupiter that led to the discovery of the first asteroids. It is often said that the law has no theoretical basis, but it does show how orbital resonance can lead to commensurability. The importance that becomes known is the sequence the Ties – Bode law saw in the number arrangement of 3; 6; 12; 24; 48; 96 etc. The incorrect application of the Titus Bode law lies in subtracting the figure of 3 from 10 leaving 7. The other way of reasoning is to add four each time to the firs value of three starting with 3 and so on. The true significance of the Titus-Bode law is that it points directly to a circular growth of 7 stages. The 7 relating to 10 is a precise derogative of the Roche limit or the Roche limit is a precise derogative of the Titius Bode principle because he two systems interlink.

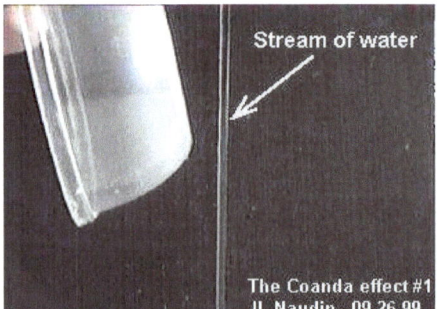

The Coanda effect

The Coanda effect applies as a gravitational phenomenon where moving liquid concentrates around the surface of round solid structures and by movement of either the liquid or the solid or both these concentrates the density of the liquid to gather and compact the flow of the liquid while remaining following the curve of the round surface. The liquid rather follows the curve of the round bowl than to fall straight to the Earth as on should expect. The liquid maintains relevance to the centre of such a round solid. I discard the idea that mass could be responsible for forming gravity because in almost four hundred years all evidence is indicating that the truth is to the contrary.

LAGRANGIAN POINT:

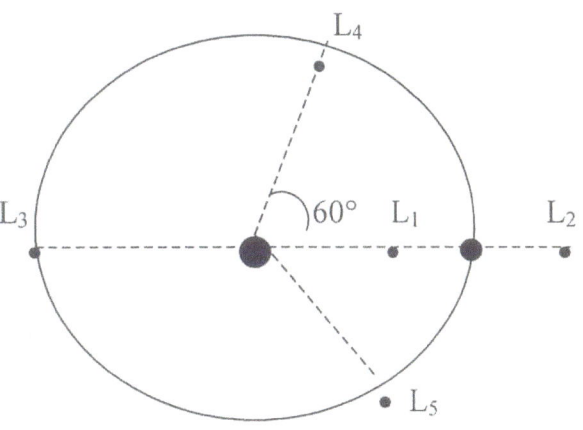

LAGRANGIAN POINT:
The Lagrangian points are five equilibrium points in the orbit of one body around another, such as a planet around the Sun

The phenomena are there and are applying! Put Newton's formula $F = G \dfrac{M_1 M_2}{r^2}$ to task and use it to explain these very common phenomena, and anyone would find it is not possible to use Newton and explain the gravity represented by this. The phenomena are there and applying so if Newton can't explain it then maybe Newton's concept of mass establishing gravity is not applying. <u>It is this last statement where Newtonian science is unwavering in their believing that mass is forming gravity which is what I strongly bring into question.</u>
Please read on to find more information concerning <u>The Absolute Relevancy of Singularity</u>

If you are a well-informed Newtonian physicist that could be taught no more and knows everything the Universe has to offer I know only nothing impresses you because you lot constructed an entire Universe you then filled with…nothing. For all others, any book that deals with gravity there are just too many and numerously wide ranging facts that form the complete picture as a whole, which leaves me unable to include a full introduction in a space as small as that which page will allow. The explaining include for instance those phenomena, which I call the four cosmic pillars, but wise as you are, you would not believe me at this point that I have cracked the coconut because I guess in your vast experience you have seen too many idle explanations in the past proving to be senseless and little impressive, therefore my mentioning my success would not matter much either way.

PO Box ?????
Some Godforsaken Town
In the
Limpopo Province
South Africa.

mailto:E-mail www.singularityrelevancy.com

P.S. J. SCHUTTE (PEET SCHUTTE)

PART 5 of The Absolute Relevancy of Singularity,

Forming part of the web site **www.singularityrelavancy.com**

AN ARTICLE SENT TO ANNELEN DER PHYSICS ON THE RELEVANCY OF SINGULARITY

Slightly adapted to adopt sketches enabling better explaining of the meaning of the text since to my shock and amazement I learned that there are intellectuals found in physics finding it hard to read and understand words explaining physics...then how were they taught?

ISBN 978-0-9802725-2-9

This article has a purpose more than anything to show when one goes into singularity the dynamics of multiplying does not exist. To produce large equations with relations that should put the fear of God running through you're veins has no purpose where singularity construes the rules of calculating. This article sets about to prove more than anything other than why when dealing with singularity the basis number applying is Π and all values applying goes by the measure of Π. Singularity puts value to space in terms of the use of Π. The atom for instance has a relevancy of displacement of $(\Pi^2 + \Pi^2)(\Pi^2\Pi)\ 3 = 1836$, which is the mass difference between the electron and the proton. In another mathematical dimension just below the atom uses this relevancy of Π to form the element table basis in which outer space forms atoms $\Pi(\Pi^2+\Pi^2+\Pi^2+\Pi+3) = 112$. The sound barrier has a displacement value of $7(3\Pi^2)(\Pi^2/2) = 1022.795$ km/h limiting gravitational movement. The sphere and therefore the Universe has a displacement value of $7/10(\Pi^6)/6$ giving the sphere a dimensional value when moving of $7/10(\Pi^6)/6$ in terms of Π. One measures singularity, not by numbers or formula, but by Π. Beware, normal mathematical applications does not apply in the Universe where singularity applies.

This shows all readers not to start looking for impressive equations because equating mathematics is useless in the Universe of 1, pre-dating the Big Bang.

Albert Einstein formulated a concept in 1905 he called **The Special Theory of Relativity** and in 1915 he introduced his assessment on the principle of **The General Theory on Relativity**. I do not quite agree with his findings. What I discovered goes far beyond the discovery that Albert Einstein formulated. I have discovered that the Universe is not employing a general relevance of singularity, but throughout the Universe there is a fixed overall state of *The Absolute Relevancy of Singularity* that is not only **controlling the Universe**, but is what the Universe **constitutes of**…it forms the Universe…it is the Universe. In his conduct Albert Einstein suggested the **Universe goes "flat "** according to his calculations. In **a "flat" Universe** all dynamics of dimensional space would disappear. That would only be the case if and when the three-dimensional space we see would become single dimensional and everything will relate to 1 because after all that is what singularity refers too to being one dimensionally. With all the genius of Albert Einstein and all the magnitude of his vast mathematical experience, he did not take the Universe further and in some quarters even become some sort of laughing stock for not proving what he proved mathematically. In this article the purpose would be firstly to prove he is correct in what he said and secondly to prove he is incorrect in his method of conduct he tried to use and thirdly to show why he was wrong in his approach to use the dynamics of formulated mathematics to try and go there into singularity.

However, notwithstanding the magnitude in significance *The Absolute Relevancy of Singularity* presents as a breakthrough in science, the influential members of the scientific establishment will not recognise my theory on **The Absolute Relevancy of Singularity**. Past encounters taught me that mainstream science in physics will again ignore the ideas that I formulated as *The Absolute Relevancy of Singularity* and I don't believe it would be well received, it will be seriously considered and much less be accepted by those with the authority to change physics principles. In spite of science fondling the idea they know every iota there is to know about science as a subject… What you are about to read is as new to the sciences of cosmology as iron was to the Stone Age people.

Because it is a new way a new way of thinking bringing truth to the science of physics I think the theory I introduce would never be accepted by the paternity governing science during my lifetime because science is fixated on Newtonian ideas, on playing games with mathematics. By creating a dream world with mathematics playing God they get into a role which makes them bent on believing in the outrageously marvellous, and the unexplainable magical powers with gravity working by mass supplying a pulling power, which is a fact never proven and accepted only on Newton's word and Newtonian cultural bias, and in the face of all of this they still claim to only use proven facts. What I ask of readers is to beforehand forfeit the culture of Newtonian bias when reading this by paying attention to what I say and not about the degree in which I stray from mainstream science's thinking. This way the exercise will present many new ideas and when explaining my new concept, the new principle will become clear. There is so much to benefit from. Science has no idea what a Black Hole is while I can prove what a Black Hole is.

I formulate mathematically what "the sound barrier" is. I prove what gravity is. By using the four cosmic phenomena, which is what the cosmos uses to form gravity, I show what "the sound barrier" is and I go much further than that. I show that gravity forms from using the **Roche limit**, the **Lagrangian system**, the **Titius Bode law** and the **Coanda effect**. I uncover these principles by placing Π within the formulating of gravity and when using Π I bring clarity to the misunderstood cosmic principles. The list of the unknowns I can then explain is almost endless. Gravity forms by movement that establishes singularity initiating a circle in using Π. I show why gravity is there, how gravity forms and what role stars play in forming gravity. There is no difference between how gravity and electricity forms and that I prove mathematically by decoding the cosmos. I prove mathematically when atoms spin they establish Π that forms the Universe. Whatever forms gravity, that has to link closely to Π since everything that has anything to do with gravity forms a circle that is Π by the value of the square radius. If mass has anything to do with generating gravity, then mass has to apply Π or otherwise mass has nothing to do with the forming of gravity. Everything using gravity forms a circle of sorts, which forms the curvature of space-time, which is Π and which curves light. The way the planets orbit the Sun and how stars spin has all to do with Π. In spinning in a circle, Π forms gravity as a centrifugal force that condenses space.

I researched the work of Kepler and found science doesn't even recognise his work, while it is his formula that forms the basis of all physics. Everyone thinks that Kepler found planets rotating, with Newton being able to explain Kepler, which makes everyone more concerned about how Newton saw Kepler's work. The formula used in physics as a principle is $F=mV^2$ which should be $F^3=mV^2$. $F^3=mV^2$ is replicating Kepler's formula in detail as $a^3=T^2k$. By using Kepler's formula we have $F^3=mV^2$ that is a precise replica of $a^3=T^2k$. The duplication is so obvious that we have (F^3 becoming a^3) while (m is **k**) and (V^2 is T^2).

Einstein also only duplicated Kepler's formula by putting $E=mC^2$, which also should read $E^3=mC^2$. Again that is precisely Kepler's formula $a^3=T^2k$. (E^3 is a^3), (m is **k**) and (C^2 is T^2). In $E^3=mC^2$ Einstein mimicked $a^3=T^2k$, Kepler's formula. (E^3 is F^3 is a^3), (m is **k**) and (C^2 is V^2 is T^2). So what is so brilliant about Einstein's formula if Kepler had it centuries before? $E^3=mC^2$ is $F^3=mV^2$ which is $a^3=T^2k$. Newton corrupted the formula when he added $4\Pi^2$ to the formula and removed **k** that Kepler introduced while $a^3=T^2k$ Newton ignored. Newton changed $a^3=T^2k$ by using the symbols G (m + m_p) to replace **k** and then declared $a^3 = T^2$.

I still wish to see the proof confirming Newton's changes as being correct notwithstanding that everyone thinks physics is entirely based on this conception. Whether the formula used is $F^3=mV^2$ or is $E^3=mC^2$, it still remains duplicating what Kepler introduced as $a^3=T^2k$. So I changed it back to Kepler's version of $a^3=T^2k$ as to better the understanding of the foundation of astrophysics and mainstream physics. The entirety of physics is not based on Newton. Physics precisely duplicates Kepler's findings while science doesn't even recognise Kepler's formula. By giving Kepler the credit due, the entire Universe becomes completely understandable…but then for my audacity to show mistakes in physics I am ignored flat! All I ever ask is prove the truthfulness of $G(Mxm) \div r^2$ because it is $F^3=mV^2$ that forms the basis of physics and that accuracy comes from Kepler's view of $a^3=T^2k$ that became Einstein's $E^3=mC^2$.

By re-implementing Kepler's full formula $a^3=T^2k$ and using Π I was able to prove what I discovered as follows:
 1) The **location, the position** and **the value** of **singularity** as a factor forming space-time
 2) Finding **space-time** by dissecting Kepler's formula in relation to **valuing singularity**
 3) Finding space-time, **proving space-time** and **aligning space-time** with **gravity**
 4) The **working principals** behind and **manifesting of gravity** as a cosmic occurrence.
 5) The **Roche limit** and explaining the resulting of a law coming about from singularity.
 6) The **Lagrangian system**, how and why that becomes the building form of the Universe.
 7) The **Titius Bode law** and I show mathematically how gravity comes about from that
 8) The **Coanda effect** and the producing of gravity through reproducing space-time
 9) The **sound barrier** by proving it **is gravity** generated **by motion** in space becoming independent motion. This I conclude because Kepler said $a^3=T^2k$ but that could also be $k=a^3/T^2$ and could be $k^{-1} = T^2/a^3$ and that is the Coanda effect. Mathematics says a sphere is $a^3 = 4/3 \, \Pi \, r^3$, which is mathematically correct. However, **Kepler said the cosmos told him a cosmic sphere is $a^3 = k \, T^2$** where that puts the cosmos in completely different mathematical dynamics altogether. There are the two distinct possibilities of a^3, which Newton saw and which Kepler saw and both are most valid, but are altogether unequal. Between Newton's $a^3 = 4/3 \, \Pi \, r^3$ and Kepler's $a^3 = k \, T^2$ concepts there is one Universal difference.

To find the invisible I had to locate singularity. I realised that my effort to locate the point holding singularity enabled me to backtrack the exploding Universe to its origins. The Universe is a sphere because it is filled with spheres filling the void spaces (not the nothings) and in that I first had to investigate the visible. Newton's mathematics says a sphere is $a^3 = 4/3\Pi r^3$ while Kepler said a sphere is $a^3 = T^2k$, and both are equally correct because the cosmos gave numbers to support its statement. Where Kepler says $a^3=T^2k$ and with mathematics saying that $a^3=4/3\Pi r^3$, we think of volumetric size of space in terms of using normal mathematic formulations. We think if it is volume then it has three sides and in the case of a sphere the measure is $a^3=4/3\Pi r^3$. Comparing $a^3=T^2k$ to $a^3=4/3\Pi r^3$ is like comparing the equal ness of a triangle and half circle and line to numerical values. $a^3=T^2k$ predates mathematics, where $a^3=T^2k$ determines positions at a period in cosmic development when only form was used going before when numbers as value were in place. It shows the half circle =180° is equal to the triangle =180° and both are equal to the straight line =180° notwithstanding the obvious differences used in form. However, the starting point of these forms has to be equal and also has to be not zero to have the end be equal and result in all being equal in value in the end.

The volume of the sphere is calculated using the formula $a^3=4\Pi(r^3)/3$. This too, has its roots in singularity.

The formula $a^3=4\Pi(r^3)/3$ is used to calculate a sphere according to mathematical physics whereas $a^3=T^2k$ forms a sphere according to astrophysics as shown by the sketch. It is at this point that physics split with astrophysics in spite of general belief about physics and astrophysics being the same. It is from the layout that the sphere uses

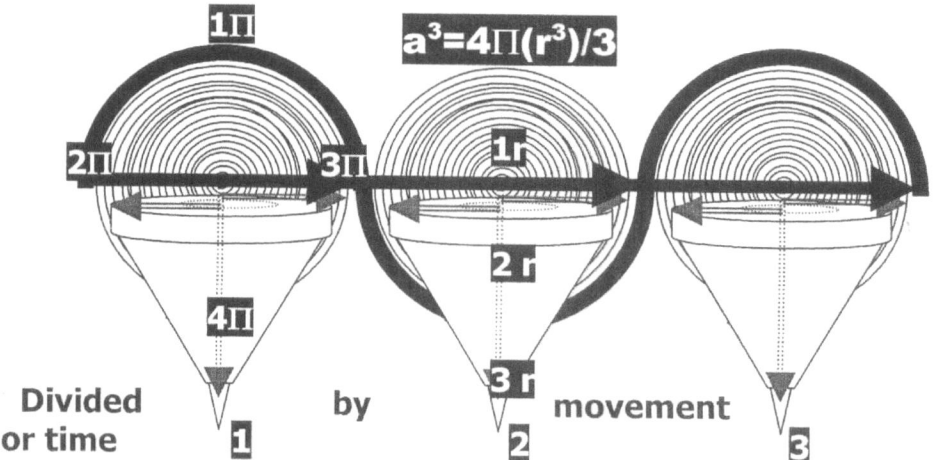

what is the basic natural form that we are able to locate singularity. When reducing the radius of the sphere the length reduces by measure as the radius becomes smaller up to a point where the radius is r^0. At that point the line that will form the radius has gone single dimensional r^0 and that is equal to 1^0, which is singularity. The position that singularity is in was always missed because $a^3 \neq T^2$ as it was always thought to be. Look at Kepler's table and see if $a^3=T^2$ as Newton said it is, or is it $a^3=T^2k$ as Kepler calculated it to be.

What Newton arrived at was not $a^3=T^2$, but instead he saw $k^0=a^3/T^2k$. In the very centre of every spinning sphere is a line forming as an axis. Then in that area all form of any possible space disappeared leaving only the dimensions of singularity 1^0. That is $k^0=a^3/T^2k$ which proves the Universe is without doubt a sphere…and we just located the centre of the Universe! The form of the sphere dictates that according to the shape at the point the axis forms the line will relinquish space. There is no space in the axis line that forms. The radius line reducing takes place in every sphere as the diameter reduces to the centre where at the centre the radius line goes single, the form relinquishes the three dimensional form the sphere has.

PLANET	PERIOD (Years) (T)	MOVEMENT (T^2)	DISTANCE	SPACE (a^3)	RATIO k
Mercury	0.241	0.058	0.39	0.059	0.983
Venus	0.615	0.378	0.728	0.381	0.992
Earth	1.000	1.000	1.000	1.000	1.000
Mars	1.881	3.54	1.524	3.54	1.000
Jupiter	11.86	140.66	5.20	140.6	1.000
Saturn	29.46	867.9	9.54	868.25	0.999
Uranus	84.008	7069	19.19	7067	1.000
Neptune	164.8	27159	30.07	27189	0.999
Pluto	248.4	61703	39.46	61443	1.004

In the above table that Kepler configured as $a^3=T^2k$ we have three distinct factors combining to form a specific value that indicates space-time $a^3=T^2k$ and moreover shows that the Universe structurally is composed in terms of **space a^3 = time T^2k** and every factor as much as a^3 and T^2 as well as k has a part and a role in forming the eventual value of **space - time $a^3=T^2k$**.

My work is forever adjudged as controversial, but what makes it controversial is the error it portrays in the truly controversial work Newtonian science not only accepts and embraces, but tries to hide from every one. It is because of the centuries-long accepted controversy in science and not this article pointing to the controversy existing. The Universe is about the way that relevancy forms valid measures in space. This is what Kepler discovered when Kepler discovered the cosmos is $a^3=(T^2k)$.. Kepler discovered relevancy where Newton missed what Kepler discovered!

KEPLER'S LAW OF PERIODS FOR THE SOLAR SYSTEM			
PLANET	SEMIMAJOR AXIS $a\,(10^{10}m)$	PERIOD T (y)	T^2/a^3 $(10^{-34}\,y^2/m^3)$
Mercury	5.79	0.241	$k^{-1} = 2.99$
Venus	10.8	0.615	$k^{-1} = 3.00$
Earth	15.0	1.00	$k^{-1} = 2.96$
Mars	22.8	1.88	$k^{-1} = 2.98$
Jupiter	77.8	11.9	$k^{-1} = 3.01$
Saturn	143	29.5	$k^{-1} = 2.98$
Uranus	287	84.0	$k^{-1} = 2.98$
Neptune	450	165	$k^{-1} = 2.99$
Pluto	590	248	$k^{-1} = 2.99$

The table above shows k forming a measured relevancy in terms of space moving around in rotation in relation with singularity where k produces the relevancy. When anything moves directionally it can't be zero. The figures presented prove that Newton's statement $a^3=(T^2k)$. is incorrect. It says $k^{=1} = a^3 \div T^2$. =3 (in Venus' case)

Kepler formulated the sphere as $a^3=(T^2k)$. Reducing the radius r to a point where r is r^0, singularity steps in filling the very centre because at one point only the form remains as Π with r being r^0. Reducing further, we find a point where even Π goes singular Π^0. At that point absolute singularity $\Pi^0 r^0$ is present but at that point so is absolute gravity present. This brings on the strength of the shape of the sphere. All cosmic objects of importance are in the form of a sphere because it is where gravity originates. In the very centre of the sphere the form dictates that the shape will relinquish all space and the form will finally be without dimension. Being without dimension means that at the extreme centre of all spheres there is a point that holds singularity because this point with no space has a mathematical position $\Pi^0=1$ although it is invisible since there are no sides to such a point to give that point any dimensions.

It is true that when measuring the sphere, Newton's method or formula $a^3 = 4/3 \Pi r^3$ is used in calculating, but **Kepler received his code of calculation $a^3 = T^2 k$ from a very high authority,** which **is none other than the Universe** and therefore Newton can't discard **k**. Kepler saw singularity forming relevancies and Newton knew nothing about that. It is the duty of the cosmologist not to reject Kepler's findings, or as Newton did, try to transform it into something that Newton could understand, because it then strays from the original meaning…but science should dutifully search for the meaning as Kepler received the formula $a^3 = T^2 k$ from the cosmos.

We can test any of the following symbolic values in the mathematical expression and also test the principal behind the expression in which Kepler stated them. By such testing $a^3 = T^2 k$ repeatedly we find that the translations of Kepler's formula into English never required any corrections in translation because Kepler never presented it incorrectly. By taking the formula on face value it can change as follows: $a^3 = T^2 k$ can become $k = a^3 / T^2$ or become $k^{-1} = T^2/a^3$. When translating Kepler's mathematical expression into English we can see what Kepler said also could read as $k = a^3 / T^2$ where **k** is indicating one point from a centre point that is space a^3 relating to time T^2. From a centre comes space-time.

The centre **k** brings space a^3 in ratio to time T^2, which is space a^3 / time $T^2 k$. Reading this correctly can't bring any dispute…yet it does…and it's been doing it for centuries! Kepler said $a^3 = T^2 k$ and that correctly translates to a mathematical expression $k^0 = a^3 / T^2 k$ which in the English verbal statement translates that Kepler said that there is a **space a^3** which is **equal** = to the motion in **the time duration T^2** thereof between two specific points which holds a relation onto a centre k^0 where from there forms **a straight line k** that is centred on the spot where space begins from k^0 **that produces k** as well as producing the circle. Therefore that spot where the specific point is at $k^0 = a^3 / T^2 k$, that allocated spot holds k^0 at a value of having the least space there could ever form. The line **k** is centred onto a spot where space begins specifically at k^0.

This point not only produces the line **k** coming from a point k^0 but represents also the space a^3 that forms the eventual circle by the rotation of T^2. Therefore from the centre holding k^0, k^0 leads to **k** that forms the revolving space a^3, which is rotating T^2 at a distance **k** where T^2 forms the outer limit of k^0. Mathematically $a^3 = T^2 k$ will also be $k^0 = a^3 / (T^2 k)$ because $k^0 = 1$. But $k^0 = 1$ also presents the single dimension where all factors are a product of one. If anyone can locate k^0 then also that person will find singularity. That is where gravity is because gravity is strongest where space is least. Then that suggests that gravity is strongest at k^0 because there space is least. That is gravity because that is what keeps the orbiting objects in orbit but also that is what Newton completely missed when he changed Kepler's work. Newton failed to recognise gravity as the only ingredient in Kepler's formula. He admitted that he, Newton missed this because he admitted he did not know what gravity is while Kepler explicitly showed what gravity is. Gravity is what keeps the orbiting objects in rotation while orbiting. $k = a^3 / T^2$ is **distance^1** = **space ^{3}/ time^2** forming from a pivoting centre k^0. That is a cycle and moreover it is a cycle formed **by space/time**. What Kepler said is that space is a^3 **being in motion $T^2 k$**.

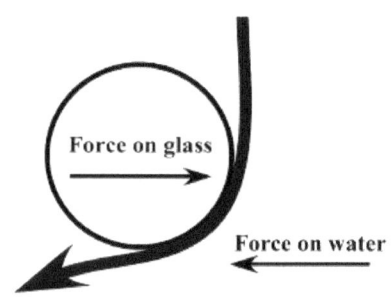

As Kepler said $a^3 = k T^2$ and therefore $k^0 = a^3 / k T^2$ and therefore we have to find k^0. As a result of examining this proposition, I located two principle positions both holding singularity. The cosmos is made up of one type (1^0) that is in two categories where one type moves and the other type does not move. The one is a liquid and the other is a solid.

The condition for the presence of this singularity that forms everything, controls everything and is everything is centralised to a centre singularity $k^0 = a^3 / (T^2 k)$ that forms by movement $T^2 = a^3 / k$ of space $a^3 = k T^2$ placed in relevancy $k = a^3 / T^2$ that is centrifugally going both ways $k^{-1} = T^2 / a^3$ thereof

(Newton's 3rd law).

This explains the Coanda effect and the Coanda effect is gravity and gravity "glues" the water to the glass by implementing Π to form singularity! *What is in the Universe is spinning*. The entirety of everything forming the Universe is spinning inside the Universe and such spinning is always in the centre of one specific point, wherever such a point might be. In the **precise middle** of all **objects in rotation** is a precise centre where this pre-designated centre is dividing the object in rotation into sectors that will **start the spinning initiation** from that centre point. This is what Kepler's formula confirms in $a^3 = T^2 k$. By spinning, the one side is coming towards while the opposing side at that time is going away. Thus, the spinning object **will have a middle point**, a very specific **centre point that does not spin** and only holds Π as a specific value because within that centre being that small, no radius can apply. We have named this position or line the axis, but the true meaning of this line has eluded us since the concept was realised. This line that forms holds no space although it directs all the space that it controls by spin. When going toward the centre where the axis forms at the very centre of rotation, the space on the one side has to end and the space at the other side has to begin with the line unable to hold space.

On the one side space turns in a completely oppositional direction from the direction in which the space spins on the other side and in between the opposing movement a line forms without the ability to contribute space. But also within the one value forming, such a line **cannot have a value of zero** because the line **is there and holds contact** to the rest of the material bringing about that **zero does not start any** line and therefore the **value of the line must be infinite**, just as described in **accordance** and by **the definition of singularity.** In dimensional terms, which I explain later on, the value of **2k** relates to **T²**. That relation extends to the next value where **T²** relates to **k**, which positions **T²**. The first space in the circle **T²** will then be located at point **k**. From the centre being in infinity, one can realise by thought that the single dimension factor is not visible, but is present all the same. Extending that into the 3D comes six **k** and any one of the six will further extend to form a seventh point as **T²**. All this forms a point that finally refers to the location of one spot holding singularity attached to space by the measure of establishing singularity $k^0 = a^3/(T^2 k) = 7$

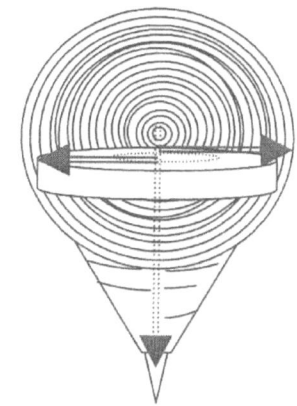

Let's find $k^0 = a^3 / (T^2 k)$ and see where it is hidden. The sphere is a circle in many facets and therefore we will approach the sphere as one multi dimensional circle. However, the sphere as such remains one circle to the power of many. When investigating a circle, one would draw a line from one edge running through a centre all the way to the other edge. In doing that we would find the measure of the diameter, which is most important when trying to establish the volumetric worth of the sphere. The circle has Π to indicate form and uses r² to establish the worth of such a circle by using the radius symbolised as r in drawing a straight line. In any circle or sphere the size only depends on the fluctuation of r in the square as a component to the circle or sphere but that does not affect the form, which comes by indication of Π in any way there may be. The conclusion from this is that no line can start at zero because that will be a mathematical impossibility.

Lines mathematically cannot start at zero because there is no evidence of zero as a factor in mathematics. Should you disagree with my statement, the question in need of answering is this: What will the length of the shortest hypothetical line imaginable be and moreover, what would the total overall length be in that case? A line or spot starting at zero would therefore be shorter than the shortest line possible. For obvious reasons can no line, or any line grow or extend from zero because such a line must then quit zero and become something, thus abandon its original value by the adding of the first value. Mathematically said it would be as follows 0+0=0 whereas if it started with something infinitively small it would be $1^0 + 1^0 = 2$ and then from using something infinitively small it will grow into something immense such as the Universe. In any circle or sphere the size only depend on the fluctuation of r in the square as a component to the circle or sphere but that does not affect the form by indication of Π in any way there may be. The conclusion from this is that no line can start at zero because that will be a mathematical impossibility. If a line started with zero, that would nullify Π ($0^2 \times Π = 0$) and that would leave the form without having any form because Π x 0 = 0.

This statement by itself excludes zero and with zero excluded one then begins to appreciate all the rest of the concepts governing corrected cosmology. If there is a distance, it holds a measured one of whatever norm or value, which is a specific length that applies and that zero or nothing then could never fill. By saying the distance constitutes of nothing we have to substitute the one factor with a factor of zero to find what mainstream says fills the Universe. Including nothing as to state the presence of that part contained by the calculation delivers the total of zero. It seems as if science has ignored this mathematical principle that 1x 0 = 0 as an issue by simply not thinking about the fact of the matter and therefore simply ignoring that which is measured forming the sole value of space. It is somehow more convenient to put the value of nothing as part of the distance in calculation because that is what is understood. Measure zero and then see how one can multiply when using zero in mathematics to reach a distance holding a value other than zero when multiplying with zero.

I agree that what is filling outer space is invisible, but also it is there, it is present and being present and there while being invisible disqualifies whatever is there from being zero because being zero will mean it is not there and we cannot deny whatever is there of being there. Then what is there will be there, while being invisibly small, but it will still be possible to form a line because every aspect of the Universe forms lines while also it will have the potential to fill space and can still form a measurable unit. That then must be 1 because while 1x1=1, 1+1=2 and that qualifies that invisible thing to be present (1+1=2) but at the same time be completely invisible (1^3=1). When realising this I knew what conclusion coming from this had to be true about that which I was looking for and that it had to be singularity because singularity can only have one value and that is 1.

I now show why singularity applies Π as a numerical basis instead of using 10 as it applies in mathematics. In singularity the base number applying to all gravity is Π relating to dimensional changes in space. This is why flabbergasting impressive mathematical calculating and inspiring equating genius has no place in singularity.

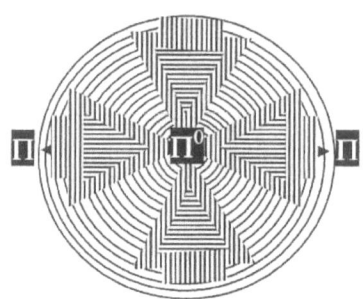

At that point is a spot where all space vacates and in the very centre only singularity 1^0 as form remains. At such a point we find the measure of the sphere being Πr^0 with $r^0=1^0$. That is where the line that represents the radius as a line disappears into singularity r^0, as it loses any place in Universal space. After even more reducing continue we get to such a point where we find only Π^0 left. At that extreme point it is where space in all form disappears, as the circle providing the sphere the form the sphere has, removes all possible form by also finally going into singularity $\Pi^0=1^0$. Going one step back to where the pint is holding Π, it is at that point where the Universe starts developing space. From Π and after Π space comes to a value as r then goes square since it then is the radius that moves when forming a circle. Mathematically the point or line can't be zero because everything is then $\Pi^0 r^0=1^0$, which is a value that is mathematically 1 and not zero as mainstream physics value the point. This puts gravity into the spin of every atom by the value of Πr^0 forming gravity and that is what Kepler's formula is all about. Kepler says that the spin $T^2=a^3/k$ dictates the gravity. This forms an entire Universe in

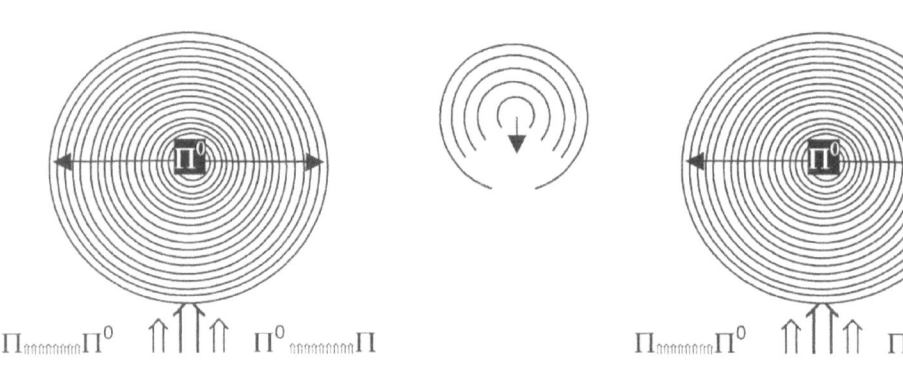

singularity without removing a Universe in space anywhere. Every sub-atomic particle known or still waiting discovery holds a centre forming singularity to the value of $\Pi^0 \Pi r^0$ and that Universe is where the Universe goes flat. By spinning, every object irrespective of size forms the Universe in singularity and every point holding $\Pi^0 \Pi r^0$ forms the beginning of the Universe at the point where the Universe starts because everything in the Universe then is equal by the measure of $\Pi^0 \Pi r^0$.

The spot $\Pi^0 \Pi r^0$ can never be within this Universe because the spot holds no space and still the spot is where everything starts that then continuous to form the Universe we all think of as being a reality. The spot (0.991) becoming the dot ($\Pi^0 \Pi r^0= 1$) controls what is in this Universe also without the dot ($\Pi^0 \Pi r^0= 1$) ever forming a part of this Universe. The spot Π^0 becomes the dot $\Pi^0 \Pi r^0$ but neither the spot nor the dot has any value in the form of space but also holds absolute validity as substance within the Universe because the spot and the dot is what could be thought of as a pre-Universe universe. It is from this continuous duplication of singularity that the Universe grows, but I am getting to that.

Such a line also connects by an angle of 90^0 to two other diameter lines where the three lines runs from top to bottom through a centre, right to left through a centre, and back to front through a centre, where all join and cross in the centre of the sphere holding not zero (0), but singularity $\Pi^0 r^0=1^0$. There are therefore three lines crossing by connecting the centre from any given point on the surface of the sphere by six points forming as one. Such points connect in total six surface points on each side of the sphere while they all support one another through the space less singularity $\Pi^0 r^0=1^0$ centre and combine the structural integrity. In that absolute space less ness in the centre holding singularity we find gravity supporting and controlling all space within the sphere as well as space connected to the sphere. That is where gravity controls and guides the space, which falls in the parameters as well as under the influence of the form of the sphere. However, in the gravity centre space goes singular meaning space becomes space less or flat. The form the sphere has allows the sphere to have a control that is coming from the centre deep inside the sphere where the space vanishes and being without space seems to keep the entire structure rigged. That is why the sphere has such strength in form and the fact that all connecting sides refer to a centre brings credence to the strength that the shape has. Any one point is integral held in structure uncompromisingly by every point holding the sphere gravitationally. How does it work in its most basic analyses? I am going to show that in this we find Π awarded a value and that Π forms gravity.

The entirety of what is, of what was and of what ever could be connects to the same value $\Pi^0 \Pi r^0$ as everything adopts the same value $\Pi^0 \Pi r^0$ when going into singularity. Where everything is $\Pi^0 \Pi r^0$ therefore the entirety is equal because the entirety is equal to $\Pi^0 \Pi r^0$ placing $\Pi^0 \Pi r^0$ in the centre of the Universe. Every time Π repositions into a newly allocated position in accordance to singularity forming the value of $\Pi^0 r^0$, then what is in the Universe goes equal in relevancy as the Universe goes "flat". However, by going "flat" this does not remove the dimensional Universe even slightly because it is a relevancy of flicker, where space is there but singularity forms prominence and then the relevancy changes to where space or better said light holds prominence keeping singularity still attaching the cosmos but forms the background in prominence. In the relevance where singularity applies the basis of numbers used to measure the basis one may use is Π.

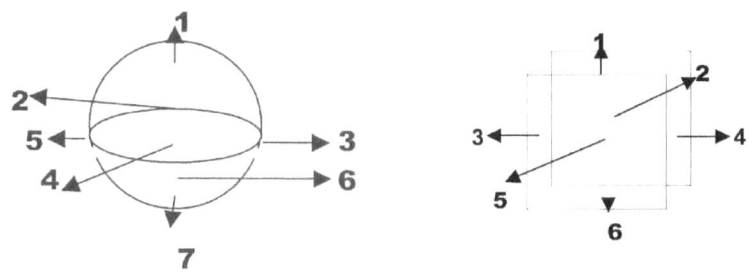

The form of the sphere dictates that contact with one point finds support of all six points acting as one across the whole structure where all six support one another as one. The six unites by singularity and the support runs through the entire sphere including the middle. Where there is no space, there must be singularity 1^0 just because the space filled with material removes zero and only material filled space is present. That means material fills the lot although the filling is in singularity 1^0 that moves in relation to a centre not moving. Mainstream science evaluates this as zero and mathematically there is no substantiating proof that it ever can be zero. If zero was a factor where all space finally halted in zero as the value, then zero would be able to remove the space from the centre and such removing would continue to remove the space until all space was removed. It will finally abolish all space in the sphere and it would remove the sphere because no space can continue from any point holding a value of zero because zero shows total absence of anything. The implication of this is that following the line down to the centre of the sphere we located the centre of the Universe. That is where gravity is. The mathematical value of such a point is $Π^0 r^0 = 1^0$ and 1^0 is singularity and could therefore never accumulate any number such as zero. That is the point where the Universe started and that is where the Universe will finally end. That is the Universe without space and only has time. It is completely ridiculous that I have to conduct such an argument about zero and still I have to.

Space can increase as it does in the case of material $k=a^3/T^2$ or compact as it does in outer space where it compresses around the rim of the star $k^{-1}=T^2/a^3$ in the area we think of as the atmosphere. But such compressing of space around every spinning star or planet would leave outer space becoming less condense overall because it removes density and it therefore will seemingly expanding. The reality about the opposing actions is that we find normal growth in material and that which Hubble first saw as the expanding outer space is space-time caused by compacting space around the margin of stars spinning by $T^2=a^3/k$ while on the rebound we find the contracting of comic space or of space-time as $k^{-1}=T^2/a^3$.

Movement in singularity is about repositioning objects by replacing objects in the relocation of objects into new space positions as time arrives at a new point on the line of time. However this movement may follow the trend set about by the Titius Bode law, which I explain in my books on sale, and the movement may incline with the Roche limit as I explain, and it may follow the exact trend singularity describes and as the Lagrangian points indicate, but all movement resonates according to and only by the law of Pythagoras and only by the law of Pythagoras. It is by the law of Pythagoras that the next instant arrive where infinity relocates every point eternity offers in relation to what infinity wants. Crossing the divide singularity holds is also using Pythagoras to do so.

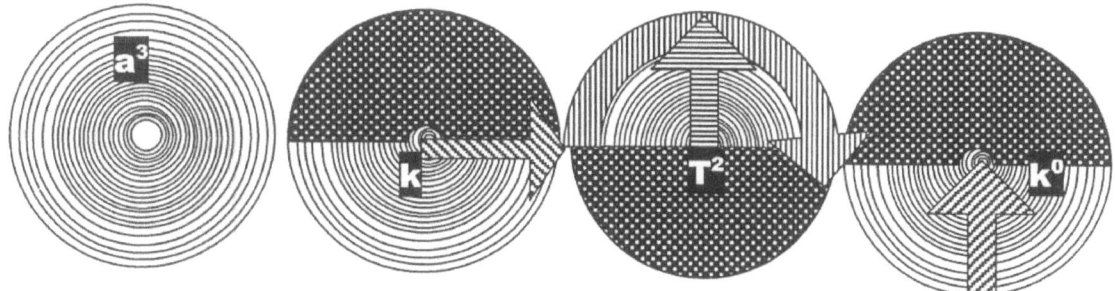

The most sensible way atoms formed is by the method of applying the Coanda principle. In the atom T^2 is almost eternal with space next to infinite because T^2 will not permit the space a^3 much room in which to be. In the relevancy we find the action and reaction of space-time flow, which is Kepler's formula $a^3=T^2 k$ and that translates to being $k=a^3/T^2$ on the one side and $k^{-1}=T^2/a^3$ in the space that becomes denser or reduces. In the times we now live in we can and do produce an optical illusion of $T^{-2}=k/a^3$, but that is implementing the use of a telescope. In the true time we find as a cosmic reality the fact of $T^{-2}=k/a^3$ is rather a mathematical statement and no more than that. In reality we have $T^2=a^3/k$ on the one side as time expands and on the other side we find $k^{-1}=T^2/a^3$.

I normally and most times only use the movement interpretation by indicating with circles how points resonate around singularity, but in truth triangles proves everything equally well.

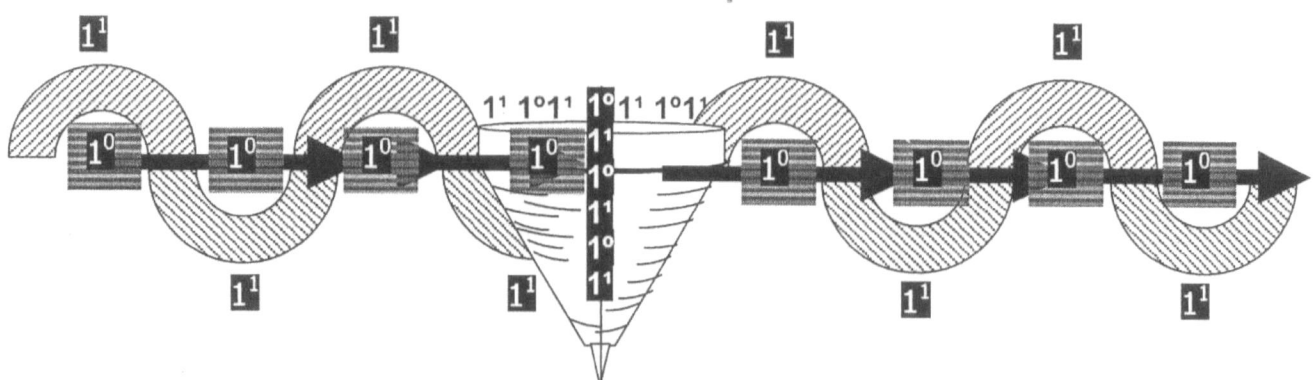

This is further proof that in singularity the line is the same as the circle, which is the same as the triangle and where singularity comes about by movement of time the lot work the same.

The proof of this we find in the way the Lagrangian system operates by implementing one line, two circles and three triangles to establish movement. This system is a direct legacy of singularity applying the law of Pythagoras into time forming eventual space. Gravity is the Lagrangian system and the Titius Bode law as well as the Roche limit forms a compliment as the Coanda effect…yes this is how time forms space as the legacy or the history of time but the movement which is time also or gravity going by another name is done by one process only and that is the law of Pythagoras. The space we see using these phenomena is the legacy which time left us as space.

Gravity forms by the measure Π leaves. There then is in this rotational movement evoking $\Pi^0\Pi$ where the turning is Π^2 on both sides of the divide $\Pi^2+\Pi^2$, which then is how I came to find the relevancy of the proton in the atomic equation of $((\Pi^2+\Pi^2)(\Pi^2\Pi)3))=1836$ and that is the displacement relation between the electron and the proton in the atom leaving the neutron with a relevancy value of $(\Pi^2\Pi)$ and from that it shows the neutron is liquid or movement and that is why the neutron shows no mass. The neutron is a conductor and not a reducer of space-time but this comes a lot later in other books. But since it involves singularity moving it calls for the law of Pythagoras to produce space. The law of Pythagoras is the triangle a^3 that is moving forward in singularity k by turning T^2. In singularity the 7 stands in for 7 points on the numerical line crossing over the line holding singularity or 1.

The condition for the presence of this singularity that forms everything, controls everything and is everything is secured in the centralised immovable centre where singularity holds $k^0=a^3/(T^2 k)$ according to Kepler's formula $a^3=(T^2k)$ indicating relevancy that forms by movement $T^2=a^3/k$ of space $a^3=kT^2$ placed in relevancy $k=a^3/T^2$ that is centrifugally going both ways $k^{-1}=T^2/a^3$ thereof **(Newton's 3rd law)**. In the line that forms when the governing singularity charges a controlling value on space that forms we have the governing singularity 1^0 that relates to the controlling singularity 1^1 but then 1^1 becomes 1^0 where it then relates to 1^1, which then becomes 1^0 that in turn relates to 1^0 that becomes 1^0 that relates to 1^1. This is eternity moving across infinity and in that creates the relation of singularity 1^01^1. This is the spot stretching to become more than the dot and become a line. This 1^0 that relates to 1^0 that becomes 1^0 that relates to 1^1 is time moving forward. Seen from above the spin will go as follows

We now find the reason why gravity holds a base in a space less ness. It is clear that gravity is in the centre of the sphere $\Pi^0r^0=1^0$ controlling from the centre through singularity everything that is outside the space less centre. That is the reason why gravity is the strongest where space is the least. We can also reason that gravity holds the sphere in form and gravity in singularity $\Pi^0r^0=1^0$ can form the sphere in the strongest ultimate shape there is. From every point on the surface of the sphere forms a diameter that connects any one point by 180^0 with the other side of the surface of the sphere using a diameter line that runs through the space less centre of the sphere.

Mass has nothing to do with falling because one see everyday on television how all things fall equal when they are dropped from aeroplanes doing stunts. The objects big and small fall as if having equal mass. When dropped from the sky a large object falls to the Earth side by side with a small object. When all things fall equal the falling proceeds in relation to the space in which they are. For centuries the pendulum arm (big and small) are used to measure time by connecting the swing of the arm to time. Mass plays no part in the process and that is the actual legacy That Galileo Galilee left the world. He proved all things fall equal and because all things fall equal one may use the pendulum wing to measure the time.

It is by buoyancy that space holds things and that removes mass as a factor while falling. Space not holding things fall with space holding things while it is therefore not the mass that causes the falling but the compressing of space which you call the atmosphere. The falling is written in the value of Π. The value of Π is $3.1416 \div 1$ or it is $21.991 \div 7$.

It is not coincidental that Π has two distinct equal values because it is due to precisely that that the first moment in the Universe came about when point 1 parted from point 2 putting space in-between eternity and infinity. There are two values forming Π as much as confirming Π. The air or space holds 21.991 when the Earth holds 7° but when spinning the earth applies the change of direction by instating the axis by the value of Π°Π which is the centre line or axis or earth centre Π° connecting singularity to the earth circle Π and then the space that was 21.991 with the roundness of the earth being 7° the space becomes Π=3.1416 or becomes the circle and the earth and the circle of the Earth becomes (7÷7)=Π° or 1 or that then aligns with centre of the earth holding singularity because the axis placed singularity Π° in centre stage. It is this changing of relevancy applying that we use to measure time because we use time to find the space it formed during the flow of time. The pendulum will not read time in outer space.

Galileo's pendulum works on the basis of 1^0 to 1^1 change 1^0 to 1^1 change 1^0 to 1^1 change 1^0 to

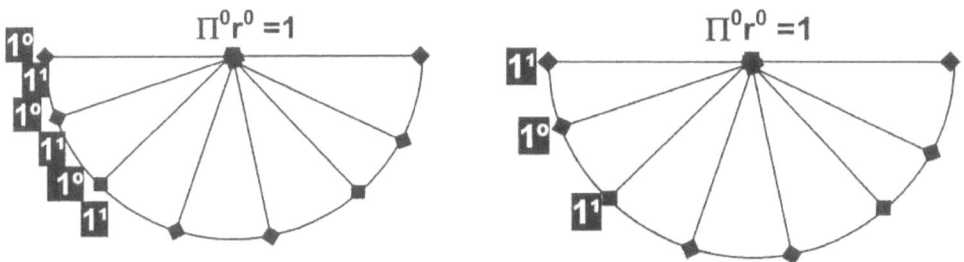

1^0 going onto 1^1 going back to 1^0 going onto 1^1 going back to 1^0

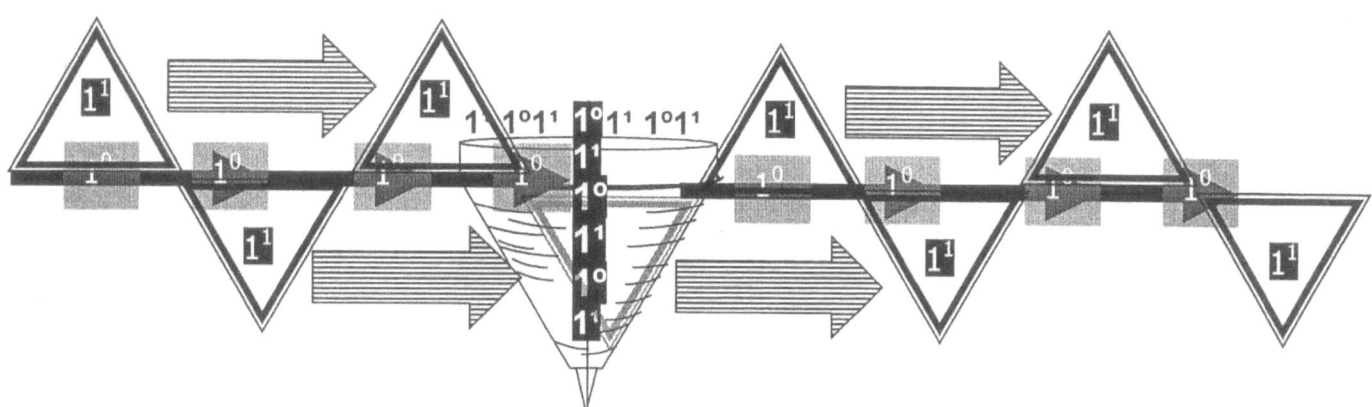

The pendulum arm takes a position of singularity **1^0** and finds the movement in singularity expanding from **1^0** going onto **1^1**, then the singularity moves space by gravity compressing and in that space moves or falls towards the earth leaving the pendulum arm to take position again in relation to **1^0** going onto **1^1** when repeating the process. The pendulum arm holds time to a measured value at the point time forms space to leave space behind as the result of time that left space. The pendulum arm resonates in sequence with time having space flow past the arm towards the centre of the earth via the circle of the earth.

In the Universe all forms are either solid spinning spheres or being outer space formed as a cube. The cube has six loosely connected sides. The sides form a weak, flat surface that connects at the corners. The flat surface produces an indifferent contact point with no special securing property features on the surface. The corners connect to only other set of corners at a time and those corners form a weak structure without any direct support coming from the other five sides. Where the cube makes contact with the sphere the cube will relent on side to the dynamic structure of the sphere. We find the Titius Bode law forming gravity by developing 7 to find a relation with 10 to form Π.

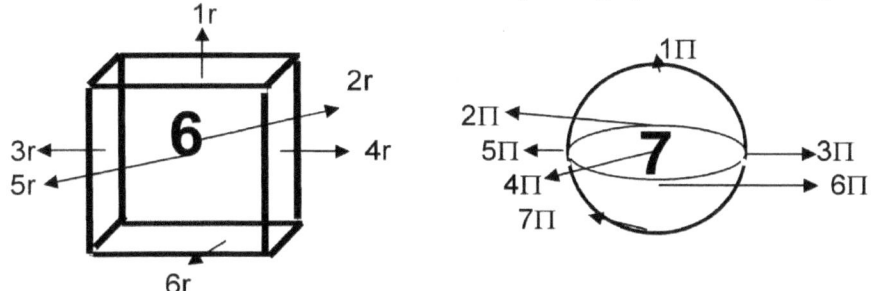

At the point of contact between the cube and the sphere the cube experiences the point of contact as losing the point to the sphere as the "bottom falls out" of the cube and without a "bottom" to support objects, the side falls to the sphere as objects does fall to the Earth. The sphere annexes the sixth point in contact with the sphere. The one side resolves its position in favour of the centre taking the side over. A body "floats" in space, but at one specific point it starts to "fall" to the Earth. That is gravity and it is a change of dimension much more than being a force.

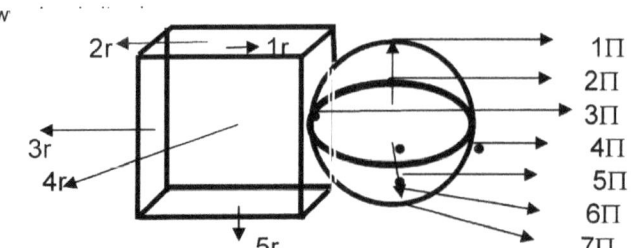

5 sides in the cube vs. 7 sides in the sphere

That is the Lagrangian system with five cosmic structures forming points holding relevancy to the centre structure where the centre structure stands in for seven positions securing singularity while the rotation brings about 7 and bringing the rotating 7 in relation to the orbiting structures standing in for five positions in space.

Kepler said a sphere is $a^3 = T^2k$, which also mathematically is $a^3 \div (T^2k) = 1 = k^0$. In honesty we have to realise that we cannot dismiss the whole formula that Kepler produced just because it doesn't match the scenario set to determine volumetric size as the Newtonian version does. Kepler's version holds a foundation based on movement and it is in the movement we find the measure and not in the size as Newton's mathematical formula does. In Kepler's formula the entire formula is formulating a circle being motion. However, with the correct interpretation we find so much more than just motion. The correct formula is $a^3 = k / T^2$: That is what Kepler brought into civilization for all time to come. He saw space a^3 being in isolation due to the time it uses to move T^2 claiming such space forming independence according to what the line k indicates. Let us look at the factors in more detail before we proceed with the rest.

Space a^3 will always be circling around as T^2 is in a position referring k to the centre k^0. That is what Kepler said when he said $a^3 = T^2 k$. Kepler indicated space a^3 will forever fight for independence and show separate individuality in remaining apart as identifiable cosmic components by means of motion. Every space will cling to independence indicated by k through fighting off the integrating of another overall unifying unit by applying the motion of T^2! The problem we have to solve is what will the cosmos use to secure such independence between all particles? What sets space apart from the rest of space? First we have to admit that Kepler was the one that introduced the following: Kepler gave us the answer to the following but no one ever took notice!
Kepler was the one who discovered **space / time** as **space a^3 = time T^2 k**
Kepler was the one who discovered **singularity** as $k^0 = a^3/T^2 k$
Kepler was the one who discovered **gravity** is holding **space-time** relative by the measure of distancing k as $k = a^3/T^2$ and $k^{-1} = T^2/a^3$

Kepler said gravity in space is about the area a^3 that would always keep equilibrium with the time T^2 it takes to travel the distance of the full circle position placed by the indicator k, therefore adjusting k as the need arrives. With k shifting in length a^3 will have to readjust and therefore T^2 will find a new relating value each time. This was the finding of Kepler and came after his intense study of orbiting planets. Translating Kepler's mathematical expression $a^3 = T^2k$ correctly to the verbal statement in English Kepler said that there is a **space a^3** which is **equal =** to the motion in the **time duration T^2** thereof between two specific points which is a straight line k that holds a relation from a centre k^0 to an end k where the two ends run from the beginning of k^0 to connect at the end of k. I might not be the smartest boy on the block but I'm not that stupid either. I know how to translate mathematics into English… and I translate as follows:

a^3 must have a volumetric interpretation because the third dimension is sure evidence of multiple conjunctions of dimensions put together in three sides opposing three sides having the third dimension in place. The fact that any symbol uses a value to the **third power a^3** indicates **space** or a volumetric established and separate unit. Using a cube by three dimensions symbolises a cube, a room, a space to be filled, a unit able to hold other ingredients on the inside when empty or partly filled. It is space because it is volume using the third dimension.

T^2 is an indication of something having a cubic nature other than the square forming motion that is provided by the motion the square indicates, which is where the moving object is representing a third dimensional object that is moving from point to point and it is this point to point that multiplies into the square. The space is moving as a unit from one point to another point and the moving between the points are represented by a flat square or following a flat distance between two points. The cubic space was in one instant in one place and then the second instant in the other and because time can never stand still or become single dimensional (this I am about to prove) insisting that time must always support the motion it consist of or space as well as time in time cannot be. It is motion that is taking time, which is motion in the second dimension moving the space in the cube.

k^1 is the symbol used to indicate a straight line between two points with a definite beginning and a specific end position. It is the location where the form in question is holding space running from where the space was to where the space will be the very next split instant that follows while time by movement repositions the allocations. This indicates points of representing k in different time positions to which the points will then be multiplying to form the square that forms between k_1 and k_2. The movement indicates not a square surface but it indicates movement by the square. This indicates the time the journey took to move the space from one point where **k** is to where **k** will be. It indicates the location of the space where from to the point where the next indication of **k** runs. T^2 will shift **k** where **k** indicates the position of the space a^3 that forms as a result of the movement T^2 of being the space a^3 indicated by the point at the end of **k**. Since time represents the square T^2 and with **k** being the distance, this fact proves that the **k** represents the distance of the ending of the space a^3, which represents the form relative to the circle that T^2 forms. It is obvious that T^2 represents the time that represents the space a^3 in the square T^2 through the motion. It is the distance moving space a^3 in the cube to complete time in duration in the square of motion T^2; therefore **k** is permitted to be in the single dimension.

<u>Let us find the smallest possible line first</u>. We have already reached the conclusion that by reducing the line, the reduced line will eventually leave all sides on the same spot on the condition that the circle spins. Such a spot must be round in form since it still holds Π as a factor next to r^0. We now are entering the domain of singularity where the visible is no longer traceable and only intellect can bring understanding of the scenario. With the line being the smallest line, such a line will start off as a dot Π that moved away from a spot $Π^0$. With all possible sides being in precisely the same spot we have all possible sides onto one spot. I chose to differentiate the dot and the spot by giving the spot a value of $Π^0$ while the dot holds Π next to r^0. Mathematically the spot is placing form evenly spread being Π coming from the single dimension $Π^0$ where the space is one (1) and holding exponentially zero (1^0).

There the space moved over to form the spot $Π^0$ and by introducing form the movement changed $Π^0$ to the dot $Πr^0$ forming a circle as a dot. Again I must draw the attention to the fact that we now are reaching into areas only the human mind can venture by understanding and seeing nothing more than with the eye of intelligence. The understanding of this concept demands our reaching the point where the mind of the animal cannot reach. If it starts with a line it then is there where that line only represents two sides being one and as such that is representing rather a flat Universe.

At the dot Π we have roundness because we have $Πr^0$ while at the spot there is not yet any round form because of $Π^0$ and only when Π being round forms, it then is requiring a shape or form and this lies beyond or before space at a point where any form of shape comes into the cosmos scenario. This part of the Universe comes in a place at a point in a location where shape and form is a part of the distant space hidden in and beyond where eternity develops.

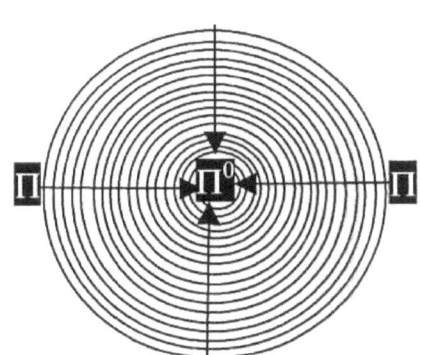

The spot is located at a point where entering the domain of the spot also at the same time is crossing the spot and landing on the other side of the spot where entering the spot is crossing the spot. Nothing can enter the allocated position the spot holds because entering the spot is crossing over to the other side of the spot. It serves us well to realise that the entire Universe was that small at a point where everything started forming because the spot that developed into the dot is still with every spinning circle...and the Universe is a multitude of spinning circles.

Gravity forms by movement that establishes singularity initiating a circle forming Π.

Following this trend of argument uncovers these principles by placing Π within the formulating of gravity and when using Π, I bring clarity to the four misunderstood cosmic principles. I show why gravity is there, how gravity forms and what role stars play in forming gravity. The entirety of what there is has to move in spin or everything falls back into singularity from where everything started. Movement drives the Universe. Everything using gravity forms a circle of sorts, which forms the curvature of space-time, which is Π and which curves light. In spinning in a circle, gravity forms Π as a centrifugal force that condenses space.

One can see from the top that singularity is established wherever spin occur. The motion generates a position of seven in relation to ten and singularity $Π^0$ by that margin manifests $Πr^0$. That means any point formed by the sphere spinning can and does start a centre in which motion forms a centre line that holds no space and of which motion forms space by and by turning around such a point then forms a line. Although everything about the sphere is in the form of a multiple circle, which results in a sphere, the sphere is not the only form present. This too has to do with singularity interpretations. We see a cube supporting the spinning sphere. The sphere can only spin within a cube and the cube can only hold a sphere. The two forms support each other.

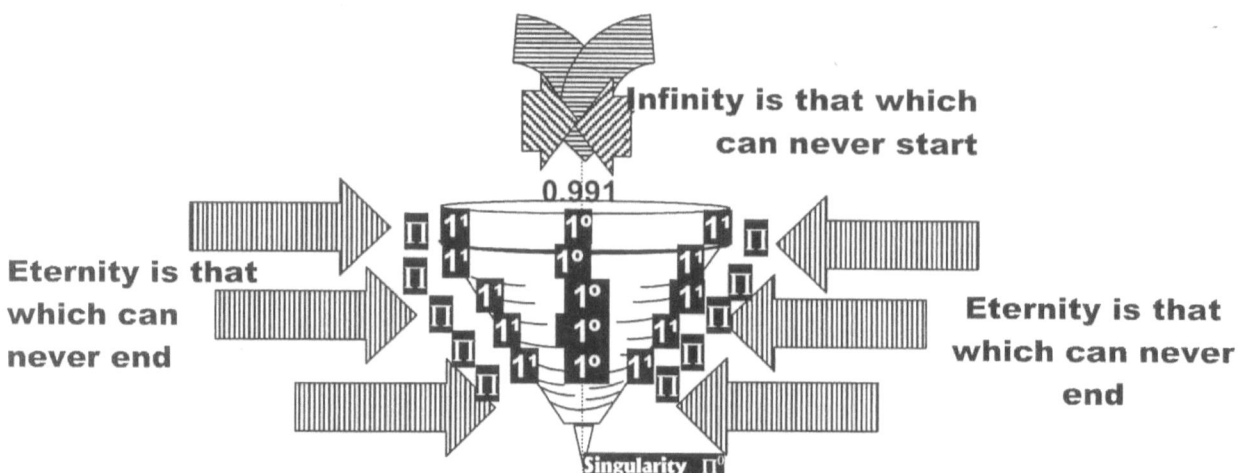

It is also very wise to remember that once anything becomes a part of the Universe, it can never leave the Universe since it then has no place to go or no gate to pass through in order to leave the Universe. With the spot becoming a dot, there must have been a time when everything in the entire Universe was that big as the spot is, and that then moved on to form the dot and in that it went on growing in relevance. The point around whichever spins becomes the centre of the Universe by singularity. In establishing such a centre containing singularity we find the reason why bullets travel more straight when they are fired circling and circling is what gives the bullet the accuracy in its trajectory that then established a cartelise singularity that establishes a value forming Π in relation to the centre singularity being 1 or as I named it as singularity Π^o.

When a rocket is fired and the spin is not present there will be no stable trajectory. The only way to secure the stability of the trajectory is to allow spin (Π^2) that enables a point holding (Π) as this will locate and establish singularity (Π^o). Establishing singularity is the most fundamental principle about gravity we can ever find. This is the one part that is most important when we go in search of gravity secured by singularity that forms the absolute relevance of everything filling Universe we have. Everything is a rotating object that holds any point allocated in Universe to form the centre of the Universe because everything in the entire Universe spins around any given point and that then forms the centre of the Universe. Every centre of every atom forms the centre of the Universe by spin! Again I indicate the precise location of such a point. What is in the Universe, is spinning and therefore what I am referring to, applies to everything holding a place in the Universe and therefore this which I mention directly links everything holding any space whatsoever in the entire Universe to one single point around which all spin.

In the **precise middle** of all **objects in rotation** is a precise centre dividing the object in sectors that will **start the spinning initiation** from that centre point. Thus, the spinning object **will have a middle point**, a very specific **centre point that does not spin** and only holds Π as a specific value because no radius can apply. But also the one value such a line **cannot have is zero** because the line **is there and holds contact** with the rest of the material bringing about that **zero does not start any** line and therefore the **value of the line must be infinite**, just as described in **accordance** and by **the definition of singularity.**

As I am introducing a very new idea, I wish to explain in better detail what I try to convey. While the toy top is spinning one will find singularity by moving the rotating line or radius progressively to the middle by reducing the length the line has from the edge to the middle. At one point all further reducing must end but the ending cannot include zero or nothing because the rest of the line is still attached to the rest of the top. As the rotating direction moves inwards, the rings will become smaller and smaller. Then we reach a point everyone thinks of as being the axis around which everything rotates. The line only forms when everything around the line spins by establishing a circle to the value of Π.

The simplicity singularity applies is shown when I show why the triangle and the straight line and the half circle are all equal to 180° and in form using mathematical dimensions this mathematical fact is bizarre. It is obvious that the triangle and the straight line and the half circle are as wide apart as the sea and the Sun is, and yet there was a period in cosmic development when the three were mathematically equal as much as they still are. In the books, not the article show using the law of Pythagoras why did Π become 21.991÷7 or then is Π=3.1416. I show using the law of Pythagoras how and why by the law of Pythagoras is a circle Πr^2 or why by the law of Pythagoras is a circle circumference Πr or $\Pi d \div 2$. I show mathematically by the law of Pythagoras why is a circle Π to begin with, but due to lack of space I can only prove it in my books as I can't prove it in the article. I show where and why did gravity start by the law of Pythagoras and what the true value of gravity is as gravity kick-started the Universe into a beginning. Can you show how the law of Pythagoras was implemented when the Universe formed, because I can show that and I do that with words since the law of Pythagoras implemented actual basic mathematics? I show why the law of Pythagoras implements the law it carries. Maybe you should try to use words one day; after all it is a helpful tool in explaining physics, because it surely helped me explain what was never explained before. I mention the law of Pythagoras because the Universe does apply the law of Pythagoras in all of the cosmos and therefore the law of Pythagoras is part of physics, don't you agree? Can you use your breathtaking mathematics without using words to tell how did it came about that the law of Pythagoras has the dynamics it portrays it has in mathematics as well as physics, because I can by using words. Use your astonishing mathematics to show why everything started by the law of Pythagoras. Mathematics can't do it because the law of Pythagoras forms mathematics and the law of Pythagoras forms mathematics as mathematics developed.

Everyone calls this line that forms the axis. Everyone knows about the axis and yet through so many thousands of years of using an axis, no person ever thought to scrutinise the principle behind the axis. Yet in all the millennia everyone was aware of the line that forms called the axis, no one took time to see it holds singularity at Π^o presenting Π. The only conclusive value singularity can have is 1 or Π^o. The axis controls all particles spinning around the line being the axis while the axis in itself forming the line represents no particles because the axis represents no space. If there was space within the axis, the space had to spin in some or other direction. Having no space would mean occupying no space which means forming no part of the Universe filled with space and yet it controls all the space as wide as the mind can imagine. Without space it does not form a part of the cosmos, but forms the cosmos as wide and as deep as the cosmos goes. The axis could not be seen but with applying intelligence the axis could be witnessed. Having no part in the cosmos in space, the axis could only be understood and never be seen. The axis could be proven but never be shown.

The axis is what controls the Universe from end to end because when there is no end there the axis provides one end to what never can have another end and the axis governs whatever spins in relation to such a line. Again I wish to press this issue to form clarity. The line forming the axis is without space and only holds form, and therefore the line represents a point not having any dimensions while it still is there without ever being there. If ever there is a concept I have to introduce, then it is the concept of how important the axis is and how science up to now missed the biggest issue that is responsible for all movement within the cosmos.

The line forming the axis is there but only intelligence will ever form the concept whereby one can realise where the line is without ever seeing the line. Anyone unable to understand this concept can never see the validity of space-time. In the axis line there is a something that is there but only intelligence can bring understanding to the understanding thereof. Only motion of space can resurrect the line coming from the point it holds as a dot.

Everything in the cosmos spins and everything that spins has to form a line that doesn't exist but yet the line controls everything that spins around this line that never can hold any space or be part of the Universe. Without having space to fill, the line can never form any viable part of what forms the cosmos, which is space.

The point in reference is the line forming the axis and the axis must be a line that never forms in space because if it did, it would have to rotate in either one of the directions space spins in and by not spinning, it has no space. **That point** albeit hypothetical, is also as much a reality none the less and is placed where that point **must be standing still** because every line **running from that point** in **opposing directions** is also **in opposing directional spin to the other or opposing side.** In considering the spinning motion in the fraction of time in the detailed instant every aspect of rotation will turn in every instant of change in time. Although the points had the same characteristics only one instant before, they oppose the characteristics it had just before and just after the very instant in which they are and to which they relate by similar points also in rotation. Looking at the graph unfold will explain my point about quarterly opposing dimensions and values unfolding.

The circle can reduce one step more when the circle eliminates r completely by returning r to a point of singularity r^0, but the elimination of r as the factor reduces the major factor to the single dimension in Π^o. That will not reduce the cosmos to zero, but it will only eliminate all potential lines r^0 to potential circles $\Pi^0\Pi r^0$ and from there the circle Πr^0 will come about by manifesting as a line but that manifesting can firstly only establish a circle Πr^2. The only value that singularity can have although the single dimension may host the entire Universe is Π^o. Pick a number and elevate it to the power of zero and in the process one may have established another point holding all points in singularity because that is the value of singularity. Only Π^0 or any other value holding one accompanied by zero as an exponential value can ever be the accurate value of singularity while singularity will then host the rest of all the possibilities in the Universe. This means that the entire Universe composes of and is made up of singularity... this much I am going to prove. Every point occupied or otherwise constitutes of singularity either under control by movement in a form we call atoms or being passive in a location we call outer space. This position one can derive from Kepler's formula $a^3 = T^2k$.

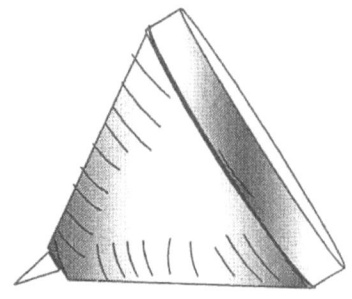

It is just a question of how to fit this sensibly into Kepler's formula $a^3 = T^2k$ and find a way that will bring much understanding to cosmology and the way that singularity connects one Universe to form cosmology. The top spinning is what connects space to form the Universe. The top being still on the ground and not spinning holds singularity at a value of the dot forming $\Pi°$ while putting the relevancy on the Earth's roundness by Π.

When the top is laying still on the earth the only movement the top has is circling with the earth. This is extremely impotent: by itself, the top can never get up to spin and if the top was a rock one Mars, the chances of it standing upright and spin was nil. The fact that the top started to spin is a contribution life made and the fact that the top resembles any form allowing it to spin is due to the contribution life made as an input on the scenario. Therefore the fact that the top adopted the spin is a phenomenon not natural to the cosmos as far as anything can move in the cosmos at large. Into this phenomenon we can read major implications and I take the reader there in the Cosmic Code. Let's not stair blindly at the top spinning and contribute the movement to be part of the free movement associated with the cosmos because if the top used the same method to spin, the top had to be many millions of time the size of the entire earth over. The top spinning is a contribution of life and not a cosmic event.

With the top lying on the ground it becomes a cosmic event but the top can never be still because even when we see the top is still it is not still as the top moves with the earth spinning around the Sun that is spinning in relation to a circle following an orbit around the centre of the Milky Way. Standing still is never possible in the entire Universe. Only securing singularity in infinity seemingly stands still but even that is not standing still because while it seems to stand still infinity is swapping relevancies to change location according to time. When the top is still on the ground the top is forming a part of the structure of the earth and doing this makes the top part of the Earth.

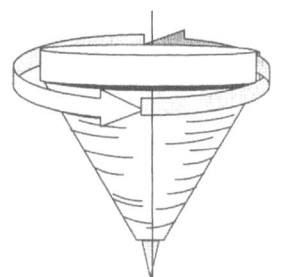

When the top spins, it receives cosmic recognition. The cosmos does not take into account why it spins or what drives the top to spin, the cosmos values the top spinning and validates the top as a cosmic entity. The fact that the top spins is due to the intervention of life acting as an energy source. The fact that life puts the movement into the top is not part of normal cosmic eventuality and life as a factor supplying movement is completely alien to the cosmos. Not distinguishing between life and the cosmos and differentiating accordingly therefore the cosmos awards the top the same status it would award a structure that just started to spin. By being able to spin, the Universe accepted by cosmic law that the top then found the ability to rise into prominence and begin a cosmic life cycle as all stars would do. With the top able to spin the earth is elevated to a controlling status but since the top does not conform to cosmic protocol the spin is not sustainable and the earth destroys the spin. This is most important to realise.

By the top spinning the cosmos now accepts it is the atoms spinning within the structure that finds the ability to have the top rotating in order to secure Π in relation to how gravity starts. In the case of the earth the atoms in the earth drive the spin of the earth to maintain control of temperature. In the case where the asteroid are, the atoms forming that particular planet was unable to maintain it spinning due to cosmic law restrictions intervening and the planets overheated and self-destruct by exploding, just as a supernova would do when the supernova explodes and left the once complete structure in fragmented debris as we now observe. This is the normal cosmic procedure, which I explain in a book I wrote to explain how the solar system formed by using the four cosmic laws applying. In it I show to the forming of the solar system while I discard the Newtonian rubbish of mass collecting dust debris to finally form the planets and the sun as it is mythical jumble.

By spinning the top no longer only serve the singularity within as it spins with the earth spinning and spinning with the earth it receives mass in terms of being $\Pi°\Pi$, but by spinning independently it forms an entity in terms of receiving a centre line holding an individual axis or centre line with three points that is evoked by the four rotating points. Now the top is no longer just one of the points confirming the earth with securing $\Pi°\Pi$ in terms of the earth, but became an entity with a gravity identity formed by the three points in singularity entertaining the four points circling and this gives the top spinning also a cosmic identity. The top is establishing its own gravity and no longer serves as mass to the Earth's gravity. Now the top validated its duel value of Π and does not only serve by forming part of the earth's solid Π. By spinning the top establishes a relation with the space surrounding the top that holds the top erect as it is supporting the top. It is in this that the top gets to become a cosmic identified object and no longer only forms a part of the earth by connecting just to the spin the earth produces.

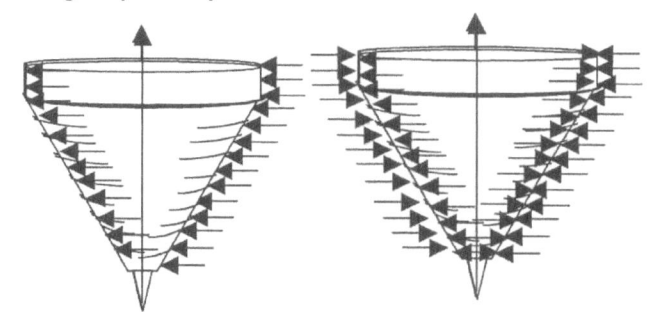

It is believed by Newtonians that the top stands erect when spinning because it exerts balance and therefore it stays erect. Well that is as Newtonian as awarding mass to get the measure of gravity. It is balance, yea sure, but if it is balance, then what has spin got to do with the process. If it is balance then the top must remain standing in balance after the spin ended. Therefore the balance they see must be connected to spin and not to balance of particles being in harmony the body of the top with the mass. By spinning in balance the balance has to refer to the compressing of the air surrounding the top while spinning. Lets inspect the procedure of the top in spin and learn from the top's actions instead of telling what the top's actions should be as Newton did. In the sphere centre is a spot that has to be there mathematically by the calculating and measuring of the defining space of any circle we find singularity $(\Pi r^2)/(\Pi r^2)=\Pi^0 r^0=1$. In order to provoke the line into action, there is motion requires to excite singularity just as Kepler indicated, where the space becomes equal to the motion and the motion is equal to the space $a^3=T^2k$.

I am going somewhat ahead of time by a cosmic mile when saying the following but there is no other way to start the explaining the process in any other more subtle way. Al movement forms by the relation there is in the cosmos between what is cosmically hot and what is cosmically cold. Movement duplicate material and therefore spin cools material down to the extent it freezes space around the object into a liquid density. To move is to duplicate material and duplicating material spreads material over a larger area distributing the cosmic heat over more space and thus movement reduces the space it claims in relation to the space it does not claim. That is why the top stands erect when spinning. By movement the top cools down it structure and in the process it cools down the space around it as well. This spinning action then "freezing" of the space around the top supports the space in which the top is and this keeps the top erect. When the gravity or spin of the top is stronger than the gravity of the Earth the top will spin. It has nothing to do with balance of any sort. When the gravity or spin of the top is weaker than the gravity of the Earth the top will fall. If it was due to balance the top will not fall when spinning to slow.

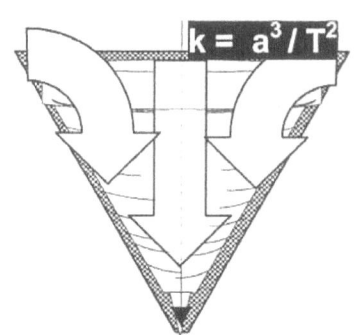

When the gravity or spin of the top is much stronger than the gravity of the Earth the top will begin to leap into the air and this too has nothing to do with balance but all has to do with relevant movement that condenses more air and the condensing of air decreases the ratio of buoyancy the top hold and this brings abut that the top will rise into the air. The top condenses the air immediately around the top in the same way as the earth condenses the atmosphere into more density. If anyone wishes to start a debate about the atmosphere being hotter, I have no space for that complex issue but read my explaining about that in one of the four books explaining the **Absolute Relevancy of Singularity.** The top spinning at speeds relative to the earth's gravity by keeping the top erect has nothing to do with balance

The gravity is in relation to the spin, which is in relation to the four points spinning which is $\Pi^2/2$ and that is the Roche limit. It is the dividing of singularity sharing space-time just as we on Earth share singularity by division between the Earth and us others moving with life and that is not part of the Earth. The total that forms from the point that spawns is seven plus five plus pi square in division of four totalling twenty one that stands related to the first seven and once again another sphere formed. However this is an eternal relevancy and as that it can never break, but also this goes much ahead of what the theme in this article would permit.

Any object in rotation will have a middle point, a very specific centre point that does not spin. That hypothetical centre line must be standing still because every line running from that point in opposing directions are also in opposing directional spin to each other. In relation to this a circle forms holding four completely opposing points to each other. While the circle of four rotates any point that had the same characteristics only seconds before, will at the new point have opposing characteristics to what it had just before and just after the very instant in which they are and to which they relate by similar points also in rotation. Due to the spinning nature of such a point with all surrounding the point every varying second, the value of such a point can only be a circle measured in Π because of its constant changing.

As the top hit the ground after being thrown in a spin when thrown with a lot of effort in the throw, it starts to move around in small circles while circling around the axis in a vigorous manner. This surging to find a new dynamic is a very important sign and is of most importance. The top is trying to reach a new dynamic by securing a further distance from the Earth centre as it tries to reduce the Roche limit affect that he Earth atmosphere has on it. It uses the gravity or movement to relocate into a new position that will grant it more freedom of movement as it then spins in less dense atmospheric air. With all the excitement of being freed from the depressing density of the Earth atmosphere and no where to take go while enjoying every minute of the excitement to be part of the cosmos as an individual entity, the extending of the drive line runs down the line forming singularity as well as from the edges all the way inwards towards the newly established governing singularity that keeps the whole top erect. This is all part of gravity Newtonian science missed for centuries by simplifying physics to be merely mathematical calculations.

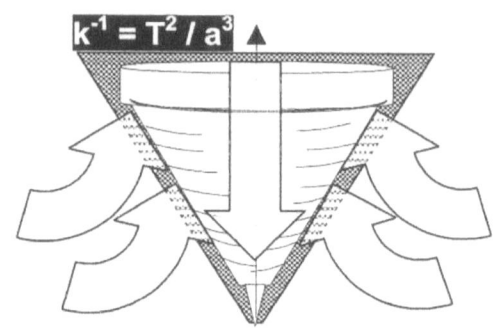

Then after losing some momentum it forms an almost motionless stance of complete blessedness as if the top is suddenly satisfied by energetically almost standing still while spinning on a precise spot while adhering to its axis and that is precisely what happens. It finds harmony in synchronising the gravity the top energises with the gravity the earth energises and the energising that goes on ids condensing the air around the top to the match the air the earth energises as the atmosphere. There is no balance except the balance between the space forming a density level that is equal to the density level of the atmosphere. As soon as the density level of the air around the top drops below the concentration of density the atmosphere has, the downwards thrust of the atmosphere will depress the top into becoming static once more.

That is why the top is spinning in the first place. The more assertive the spin is in velocity the more reaction there is from the lines forming space that is by increased density in space running towards the centre of the top and

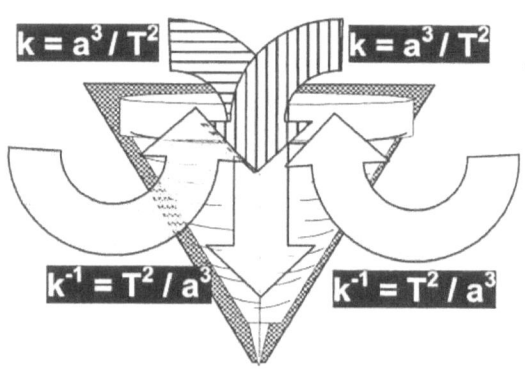

extending the flow of space inwards as it is expanding the material forming the top outwards. In real terms the space of the top expands as the spin is in contact with more space surrounding the top during the same time in period and a bigger unit of space fills the area holding space in which the top spins. In this the space in which the top spins or the space occupied by the top or the space that top holds and with the vigour and excitement of contraction of space has to allow the space holding the structure in which it is to expand as well as in order to compromise for material relevancy growth to extend the influence of the spinning material to allow for the expanding. This also serves the purpose of room to fit in the influence extended to the newly acquired singularity governing the motion that extends and asserts influence to the edge of space-time. The support that the spinning top finds in the established governing singularity ($\Pi°$) keeps the top spinning in an upright stance (Π) only supported by the singularity that takes charge of the spinning (Π^2) of space-time within the set boundaries (Π^3) to establish time within the Earth's time restraint ($\Pi^3 = \Pi^2\Pi$). What keeps the to erect is the very same principle as what keeps the Coanda effect going. The space around f the top is liquefied from the state of gas it was when being in outer space.

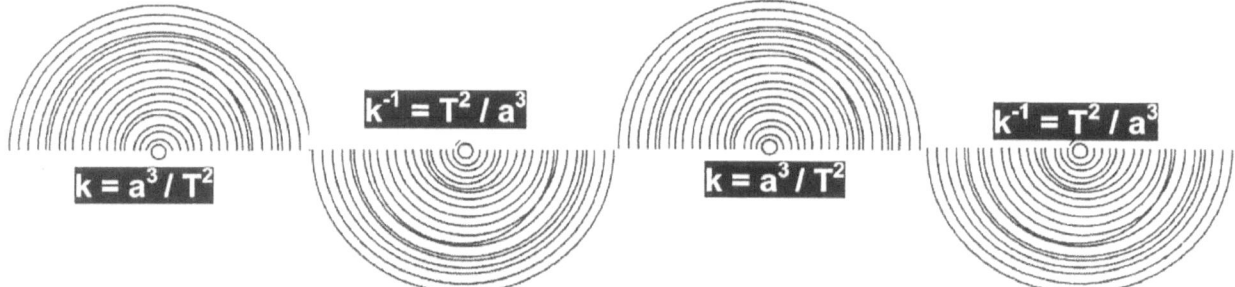

The heat differentiation between what spins and what spins not that should supposedly under normal cosmic law drive the spinning top, will apply a drive that come from the governing singularity accumulating the heat in concentration by the contraction or cooling ability the top singularity acquired. This is going miles ahead of what this article aims to prove but I prove this statement over and over in my books with much information in very detailed arguments. But in this case the spin is a result of life's ability to manipulate space-time and lead cosmic events. The heat that would establish such a drive in motion in real cosmic terms would require a lot of nourishing and sustaining from a large number of spinning atoms maintaining a low temperature that produces a large flow of space-time contracting towards the object spinning. By creating a pole that is hot and a pole that is cold we have the same movement there is when a drop of water falls on a red hot plate. It runs about while boiling the water into vapour. In this case it spins around turning cosmic space vapour into liquid we cal the atmosphere.

When spinning veraciously with sufficient energy the top gets into a fighting mood making the top very reluctant to give up this newly established cosmic freedom. The behaviour now attributed to the top is normally the manner how a star develops in the galactica cocoon and how the fledgling star gains its birthright to leave the nest of the cradle of the galactica but I get into more detail in the **Absolute Relevancy of Singularity: The Theses**.

In real life this start of movement comes from the atoms forming a sum total of energy evoking a centralised axis and he formation starts to form the unit that starts to spin around the newly formed axis that can support the generating the required gravity to secure the heat that would unleash such a drive. It forms singularity by connecting to the centre singularity and that singularity is the singularity that is governing the structure movement. When this happens in cosmic terms within a galactica the object rotating comes to life and release the new star from the blanket of heat that covered the star up to the time of this release from the galactica centre of the galactica.

If any person wishes to echo Newtonian sentiment that planets from by using $F = G \dfrac{M_1 M_2}{r^2}$ to collect tiny dust and compress the dust into particles that eventually become rock that then compress further to become complete structure and form planets. Please, please, please do the calculation how this mathematically happens so that I can learn because this is part of the biggest scam that was ever criminally conducted in the human history of all times and the perpetrator was Isaac Newton. If you wish to defend the honour of Newton use $F = G \dfrac{M_1 M_2}{r^2}$ and prove it is possible to collect debris and dust and by mass shrinking the distance of the debris being apart, compress that lot of dust into a star. I was told time and time again that physics is done in a manner by using mathematics to prove facts and what I do is invalid. Show me physics employing mathematics and prove how Newton sys the cosmos forms. No one ever could prove what I ask so lets get on with reality!

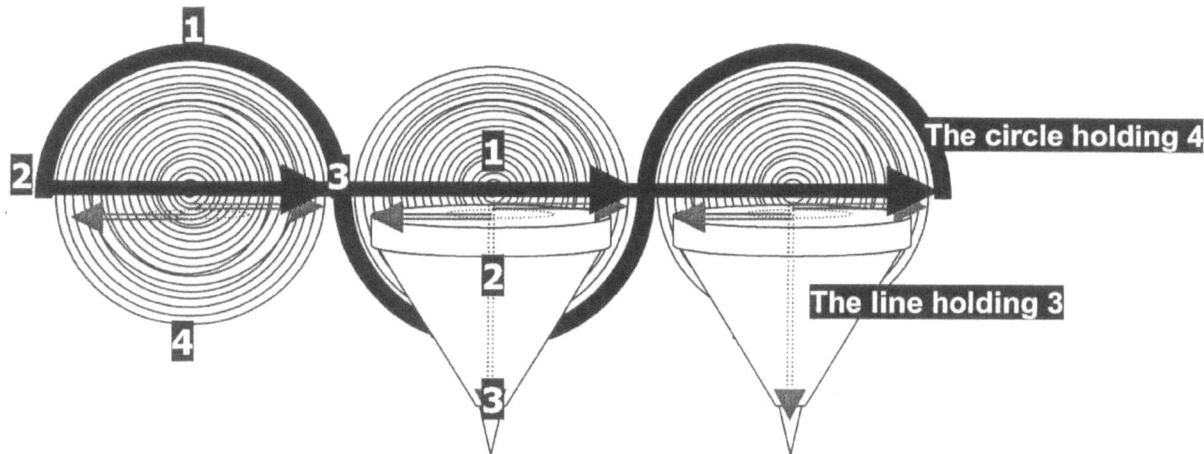

The balance is a control of motion that is established as a flow of space-time supports the ends (4) holding time while this generates the space (3) singularity containing and creating the space (3) in which the spinning takes place. There is a something (if you wish I'll use the term force reluctantly and although I strongly hesitate to use such an outrageous term) that is generating power to keep the top upright while the top is spinning.

The example we can gather from the top shows how desperate a governing singularity can become in desperation to stay active and how such an exited singularity can put up a fight for sustaining movement and independence. Just before toppling the top is in a fight for independence while the Earth is restraining the independence and we witness the fight going on. The fight goes on until the Earth finally suppresses the last bit of motion that the top had and the top uses the last motion it has to defy the Earth's domineering control.

Let's quickly establish events as they translate singularity from a dot to a controlling entity serving singularity that is commanding space-time through the establishing of a separate individual drive. The motion comes about which proves to be that which generates the gravity that drives the individuality in the top. By the motion and the singularity the top evoke a graph forms where the graph runs along the line of time. This makes nonsense of Newton's presumptions that the spin nullifies the space. When the motion exceeds the level of the Earth gravity, the top shows an eagerness to rise to higher levels of independence in the same manner that an electron reaches into higher rings of energy in the atom because the top with motion is in an electron relation with the Earth filling the role that the proton would play its role within the atom and that puts the atmosphere in the neutron role.

The energy that is charged, has the dynamics to stand its ground against the might of the gravity of the Earth that is under normal circumstances controlling the stance that the top has to take, but as if inspired, as the top seems to be revivifying by motion, the top is fighting and rebelling against the Earth's constraining gravity. The top is self-driven, as an electric motor would be. The difference between it and an electric motor would be the origin of the source from where the energy comes which drives the spinning top.

The top stands upright as individual as any self-propelled object can be. Although gravity is retaining the motion of the top, it is not contradicting the motion, all though it still restrains the actions. The earth's gravity is not combating, but is merely suppressing the motion. In this there is no trace of evidence linking mass as a factor to any of the above-mentioned actions. What we would think of as air restriction is no restriction because from the restriction comes support that keeps the top standing on a very thin needle edge. The top should tell us so much about nature if we would only listen and learn and not tell nature what we think nature should tell us.

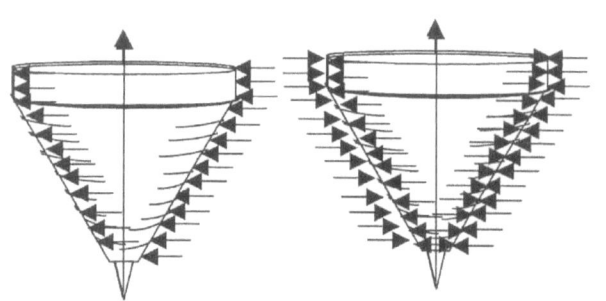

What happens is that the space condenses (21.991÷7) by the turning of the planet or star and the compacting of space surrounding the spinning sphere results from the rotational movement of the earth that brings about that this compresses the space of air or atmosphere (21.991) into more density (÷7) which is done by movement of the space surrounding the turning object (21.991÷7) albeit a planet or a star or even a spinning top and moves space filled with whatever or unfilled, going vertically down towards the roundness of the Earth or (Π=3.1416÷1). Everything within the concentrating space will come closer to the surface because it is the space that moves down to the earth and not only the object filling space, but everything within the space including the space that falls downward.

The falling body never stops falling but find that mass comes about when relevancies changes from Π=21.991÷7 to Π=3.1416÷Π^0. By touching the Earth, and by that ending the relevancy of Π=21.991÷7, the object then becomes part of the earth circle Π=3.1416÷Π^0 and having contact with the axis it becomes part of the Earth singularity distribution and only then finds in this relevancy applying the reward of mass.

The body never stops falling but as the earth by density restrict the body movement vertically according to density, the falling becomes a tendency to move downwards in order to unite with singularity formed within the centre of the spinning earth. If you want to get childish you may say gravity is a force of pushing down to create mass but nothing is pulling anything anywhere. This is all about relevancies changing and relevancies reapplying positional changes, which is what gravity or time is. This is how the Universe goes flat or singular. It is It takes space and connect it to singularity by circling around the axis $\Pi°\Pi$ = Π^3 / Π^2.

The spin of the sphere constitute of a change in direction to the value of 7°. In the centre of the circle the axis are Π^0 and therefore the spin makes the directional change reform to singularity or change to the value of the circle in relation or in relativity with the axis 3.1416÷Π^0 and the circle, which is space, goes singular or goes flat bringing about the much argued flat Universe. This is how gravity puts multi-dimensional space going into singularity or Π^0. This means that the 7° becomes one or singular.

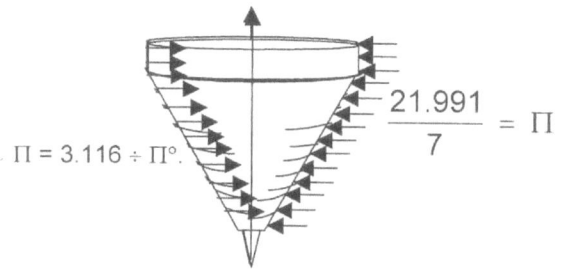

On the top of the equation the value of Π is 21.991, and by revaluating 7 to become 1 that value changes to the value of 3.1416 being in relation to Π^0=1. The rest of the explanation that will bring proof to my statement is far too bulky to offer it at this point but I did say it in the article you refuse to publish because you refuse to read it. The space reduces (21.991÷7) to conform to singularity 3.1416÷Π^0 by the rotation of the sphere that produces an axis by initiating singularity Π^0. In the books I show the very reason why is Π=21.991÷7 and I use the law of Pythagoras underwriting the Titius Bode law, that conjuncts with the Roche limit as well as the Lagrangian points to prove the Coanda effect and the Coanda effect, as I show in the article you didn't read and therefore wouldn't publish, is gravity by principal.

When the top spins the relevancy changes to the line from forming as a dot Π^0 becoming a line Π. The line Π forms as a result of the top forming space Π^3, which is in place as a result of the movement that the top acquires Π^2. It is singularity without space so being a line or a dot makes no difference. The top no longer holds only a dot Π^0 in the centre, but generates the relevance Π by forming $\Pi°\Pi°$. The top, by moving adjusts Π to form space by movement which is $\Pi = \Pi^3 \div \Pi^2$. All of this is what makes gravity be what it is and all of that Newton missed and Newtonians never saw since all of that is covered by a blanket called mass being responsible for gravity.

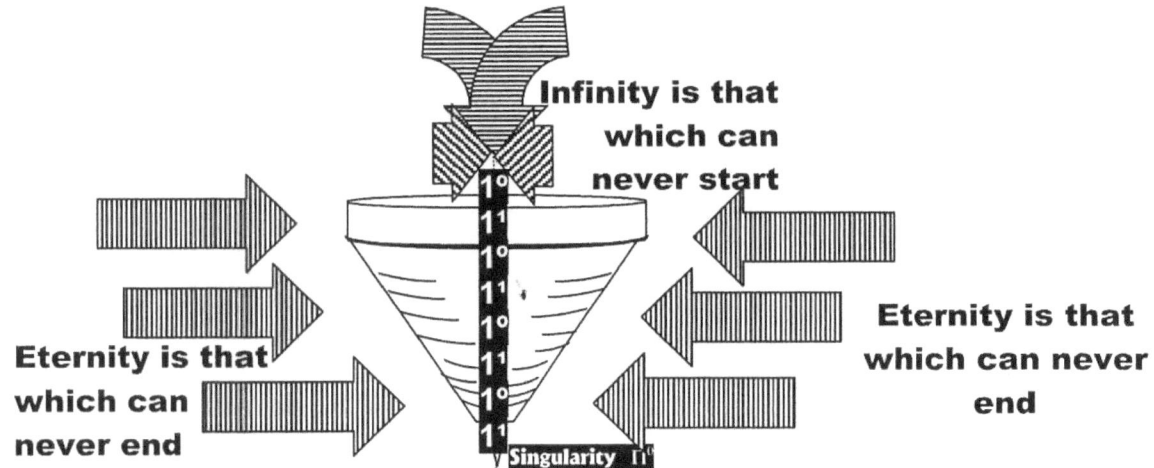

We know that dividing the radius into the circumference produces Π. We know that Πx the radius is the circumference, so $\Pi x r^0$ = the circumference and when the circle is at its smallest the circumference is Π. But never has anyone gone further and asked why is the smallest circumference $\Pi x r^0$. I have the answer and the answer is also the answer to gravity and explains what gravity is. Forming Πr^0 leaves the ratio where every 21.991 lines or dots, the line will bend by 7. Every 21.991 the ratio will reduce by 7. There will forever be an inclination of 7 for every 21.991. That is the formula that produces the circle, but what brings this formula about? What would bring about the 7 as a factor because the 7 are very pertinent in the entire relation of Πr^0? We have to look at the top spinning.

This relevancy is so pertinent that from this we can surmise how the Universe came about and what happened the very first instant the Universe came in place. The Universe introduced space by introducing the four cosmic laws named **The Lagrangian system,** 2) **The Roche limit** 3) **The Titius Bode law** 4) **The Coanda affect.** The phenomena never made any sense in the past, bust once one attaches gravity to their meaning in the correct manner by implementing gravity as Π, the true function of the phenomena as far as implementing gravity comes in place. Then one can clearly see how gravity forms Π to put an entire Universe in place.

The circle forming Π uses 7 to indicate the roundness of the circle but the 7 holds its roots deep within creation. It indicates how the Universe started because this is the way a star will start moving and it shows how as the infant star starts generating gravity just as the top starts to spin when it is thrown by life. Life can create nothing and that is true but life can mimic all laws in the Universe. Time is eternal movement and will be with us always. The line in infinity is still present while not being a part of the Universe. This line is always ready to be in place when the slightest movement orders it in place. Before the Universe was in place eternity and infinity was in perfect harmony and the line forming singularity validates this fact. Before infinity parted from eternity, eternity met infinity on one spot as eternity came from the past (1) forming the present (2) to go onto the future (3) but also returned to come from the past which was the spot held by the future and this we find in the fact that the line forms 1 when not spinning but as soon as it evokes by spin, 3 points form even now. Then heat and cold differentiated values and space landed in between eternity and infinity. As eternity moved in relation to infinity but not forming a part of infinity any longer, eternity had to follow a path by never going away from infinity (3) and always returning to the point infinity holds but never lash onto the point again. With space parting the points, eternity had two points (the past and the future) before the partition came about and infinity held both the past and the future while infinity had the present as it still gas presently. By eternity also moving, the two points it held opposed each other (the past and the future) and since it moves, by the movement it became the square of the two because movement is the square and not a flat blanket-like surface with squares embroidered on it as Newtonian science depicts it by using grand mathematics to understand singularity.

Then we had two point holding eternity in place going square by movement to form 4 points serving eternity and infinity captured the first three points held by both and since eternity could not release from the two it had but had to duplicate what it had, eternity by movement became a circle captured by the line. With four points captured by the line of three points the circle coming about is eternally returning to infinity but never complying with infinity because if mismatching temperature or movement (3 against four). Material will always be colder than outer space. It is because material spin and outer space moves by expanding due to overheating. This is where I start when I start to explain the first moment but I use a shipload more information to do explaining when I explain the star in the book I do so. I involve the four cosmic pillars to substantiate the claims I make because all four still work the very same way as it did at the beginning of the Universe. The three points serving one part of singularity combined with the four points serving singularity unites as seven to form a circle of either 3.1416 or 21.991÷7. The seven going to one is eternity matching infinity by movement. But since seven moves it is seven that has to produce gravity. How do I know all these facts, because we can see from the top it is still doing what it did the very first second.

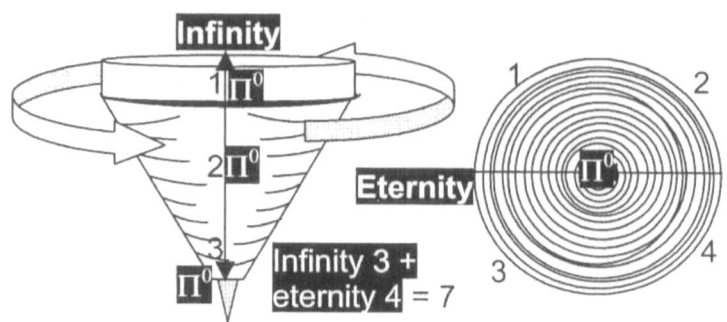

When time started infinity as well as eternity had altogether 3 positions, the past, the present and the future. It is still forming the very line in the centre of the top as it forms all lines in the centre of all things spinning. Then eternity parted from infinity when heat separated what is cold from what is hot and eternity formed one more point than before when it had the three points. With infinity and eternity then jointly having 7 the cosmos came into rotation. In the aftermath post big Bang we now see the phase of cosmic development where the tow sectors try to unite and this brings along the contraction. When Π forms it does so on the grounds that 7 rotates. The circle forms by a change in direction by 7°. Every circle has opposing sides forming in relation to the axis line. If the topside goes rite then the bottom side has to the left. If the rite side goes down then the left side goes up. There is this double presence of a change in direction forming on both sides of the circle. The 7° move and by moving 7° goes square 7^2 and that is Pythagoras.

Gravity is about the reducing of space to singularity. In spinning the sphere contracts by measure of 21.991 reducing to 3.1416 while 7 is reducing to form singularity, but also gravity forms when the 7 comes from the past to the present 7 and onto the future 7 and this became 21. Not only that but with singularity advancing from infinity to become one it proves that even as we see singularity as one, singularity also is multi dimensional but that ability is beyond our scope we have being in the Universe. The dimensional change that Π undergoes shows that singularity repeat into a new location by the value of 0.1416 and then as the new 7° as a redirection forms as at first becoming 0.991 that then progresses to 1. That is how the cosmos started. Infinity holding eternity on one spot coming from the past to the present being one spot and onto the future being one spot the cosmos was singular monotonously eternally by repeat.

That is gravity reducing 10 to 7 = 1.42857142857142857142857142857714

On both opposing sides the turn redirects by 7° and that is (7+7)÷1.42857142857142857142857142857=0.98
While Π = 3.14159265358979323846264338327950

I am not using all those decimal points every time I am pointing to a value.

The perfect became imperfect as temperature differentiation brought a partition between infinity having 3 points and eternity always returning imperfectly with 4 points and a cosmos landed between what is a line (3points) and a circle (4 points). If science wishes to find the origins of life they will have to start here because it is here that life still is vested within the Universe…and good luck with that effort because looking at life at that point goes far beyond the ability of the flat minded Newtonian. Π has a value of 3.1416x7=21.9912, which shows the margin with which time allows space to grow. Time is 3 or in relation to 7, it is 21 and time extends Π to a margin of 0.1416 or in relation to the spin of 7 then 0.991. It is this extending of time 0.991 forming space 1 where life is.

Movement is time, not only is it time related but movement is time. Whether the movement is electricity or wind blowing or water flowing or just mirages on sand dunes imitating water, it remains time discrepancies applying. Time following the line which I refer to as the spot 1° forming the dot 1^1 is according to all accounts delivered by Π the turning of (1^1going(7) becoming 1° going (7) becoming +1^1 =(7)) going(7)becoming 1^0 forming (7)x3 =21

1^1 Time coming from the past
+ 1 Time in the present
+ 1^1 Time going onto the future
= 3 Value of time moving

$1^1 + 1^0 + 1^1 = 3$ and $3 \times 7 = 21$, which leaves a value of 0.1416, unaccounted for.
$\Pi = 3.1416 - 3 = 0.1416$

Movement of material through space is $(7+7)/10 = 1.4$

$1.4/0.1416 = 9.86 = \Pi^2$ and that proves that time moves by replacing on e position in the future that then grows to form one to become three and to result in one more future point.

There is a Universe going on we know nothing of. This Universe runs in the dimension of singularity and it is there we will find life. The Universe we think of in terms of singularity

It is the point forming the very centre that plays the part as the <u>controlling singularity</u> within the Universe I have named as **Infinity,** which is better known as the axis. It is where nothing can go smaller and anything within that point can never reduce. That point is where the entirety called the Universe begins and where everything holding substance begins. Once one accepts the fact of singularity being present in that location, that accepting of singularity then is contradicting all the things we know and we can measure and we recognise that point being present by merit of the fact that the point referred to is not being formed by any of the things we can recognise. It is made up of everything we don't know and constitutes of everything we are unable to recognise or visualise. In that spot there is no space. That spot holds **Infinity.** In that space there can be no motion because there can be no space to have the motion within.

It is formed as a line that is so small that our human reality by perception declare that point as not being there and the only reason why we know it is there is because of the results it left as an imprint of its not being there. We cannot detect it but notwithstanding our failure to note it we can recognise the dot on the merits of its absence and while in our Universe it is always absent, reality disallows the dot ever to be absent, because it is never absent. It cannot be absent. It cannot go absent but it can never be there where it should be in a place from where the third dimension forms and it is always present if I wish to locate it. It is **infinity** that can never go away.

Infinity is ever present and eternity will never disappear. Infinity can be provoked just by allowing spin to form. The line forms with three points forming the line and the line forms as a result of four opposing points forming a circle $\Pi = 3.1416$ in relation to the line. That is what Kepler's formula shows. A circle $T^2 = a^3 \div k$ stands related to form space forming.

I named the other part of singularity forming space **eternity** because that area never become bigger, or become more or find an end to the outside. Whatever was and is and will ever be is locked in that space I named **eternity** and it is **eternity** that never ends because **eternity** can never end moving. What we think of, as expanding is never ending movement giving eternity the eternal motion that will go on forever. The "so called expanding" of the Universe $T^2 = a^3 \div k$ is where singularity is shifting relevance **k** from liquid $k^{-1} = T^2 \div a^3$ to solid formulated as $k = a^3 \div T^2$ and the process whereby this happens is precisely the same as the Coanda effect. Getting back to my first argument about a line and that no line can start at zero but has to use singularity as a starting point, this is all the proof I require. The line **k** coming from the centre (singularity k^0) forms an initial spot Π^0 becoming the dot Πr^0. However, I went on to say that the line used to start with has to continue in order to repeat the same that began the line.

Therefore the line started with Π^0 and it has to continue with Π^0 until such a point, as it must end with Π. Whether the line is Π^0 or is r^0, or uses 1^0 the outcome all refers to singularity being used. By reducing the line we come to the end represented to the end of the mathematical equation of the circle but the circle does not singularity where end there. That is what Newton did not recognise from the figures the cosmos sectors. By Kepler. The circle only secures the final cosmic figure and the value to the space Π^3. all things have equal value. The movement of the circle splits singularity in two same. Kepler said forming Π the circle has to form Π^2 due to the movement coming about in securing Kepler chose to use different symbols to those being valid, but the concept remains the that $a^3 = T^2 k$ while I show that $\Pi^3 = \Pi^2 \Pi$. It still confirms that movement $\Pi^2 =$ is the forming of space by three dimensions Π^3 in relation with the movement Π^2 being relevant Π to singularity Π^0.

Gravity is the contraction of space density taking Π from a value of $\frac{21.991}{7} = \Pi$, which is what is in space to the rim of the earth, which is $\Pi = 3.116 \div \Pi°$. This indicates contraction by the earth's change in direction by 7° to alter the relevancy applying from $\frac{21.991}{7} = \Pi$ to form $\Pi = 3.116 \div \Pi°$.

$\frac{21.991}{7} = \Pi$

$7 \div 7 = \Pi°$

$\Pi = 3.116 \div \Pi°$

When we look at any sphere Π^3, for instance the Earth, turning Π^2 we find the circle wishes to move straight following the line by repositioning Π as it would do by time going singular. But by turning $\Pi^2 = \Pi^3 / \Pi$ it places the axis line or centre line which is singularity to the value of $\Pi^0 = 1$ in place.

The line has two opposing sides turning directionally against each other while turning with each other. By moving or turning this involves time duplicating space by the square Π^2 on both sides of the divide $\Pi^2 + \Pi^2$ and using the same divide or the same axis or the same point serving singularity we have 7^0 crossing the same point in singularity Π^0 and is therefore connected by singularity $\Pi^2 = \Pi^3 / \Pi$. There then is in this rotational movement 7^0 standing in for Π^2 on both sides of the divide $\Pi^2 + \Pi^2$, which then is 7^2 on both sides of the divide $7^2 + 7^2$. But since it involves singularity moving it calls for the law of Pythagoras to produce space. The law of Pythagoras is the triangle a^3 that is moving forward in singularity k by turning T^2. In singularity the 7 stands in for 7 points on the numerical line crossing over the line holding singularity or 1. By moving 7 has to go square T^2 and that means 7 goes square 7^2 twice $7^2 + 7^2$ crossing the same divide $\Pi^0 = 1$. Since all movement in singularity has to enforce the law of Pythagoras we have two triangles holding 7 dots moving across singularity. I don't want to get too involved by bringing in numerical outlays because then this can truly become complex.

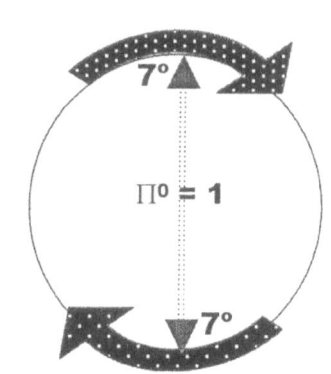

The circle spins in duel directions. On the one side it would go left if on the other side it would go rite. The one side hold a directional change in singularity by 90°. As it is going sideways it changes to going down. This produces a rite angle triangle of 90° and in it the law of Pythagoras produces direction changes.

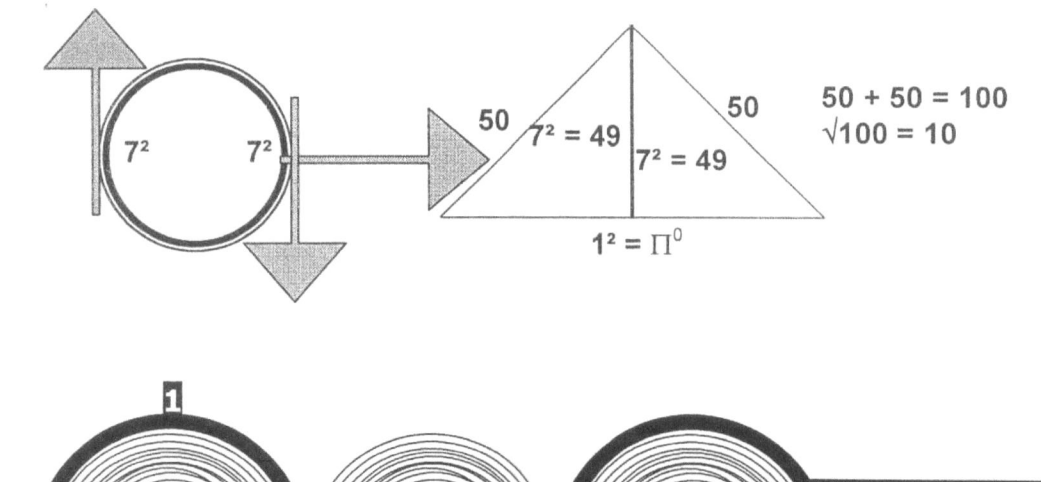

The circle spins in duel directions. On the one side it would go left if on the other side it would go rite. The one side hold a directional change in singularity by 90°. As it is going sideways it changes to going down. This produces a rite angle triangle of 90° and in it the law of Pythagoras produces direction changes.

Since the square of the turn of the circle places by the spin and the direction change we have 7 holding a relation to 10 in space because it is space that has to carry the value of 10 when material circles by 7. There is a connection between space surrounding the spherical circle turning and the sphere. Te circle holds the value of 7 as in 7° and this we find from looking at singularity controlling the circle by movement

Time runs on a line and even where we see space develop when using a familiar formula, we directly associate with what is used to measure space to time forming the ingredient applying as singularity. A sphere is measured using the formula $a^3 = 4/3\Pi r^3$ and that is correct, but the formula finds roots in singularity and moreover in the line time leaves space to form. Looking at the formula used to determine the volumetric size of a sphere we trace the circle holding four points in singularity relative relaying to three points within the line forming as the axis and this is also applying by connecting the circle to the line in relevancy or mathematically pronounced as dividing. The circle never separates from the line and never does the two factors part from Π. The circle rotates around the line and this rotation around the line secures time forming space and time forming space produces what we think to be space-time. That is how gravity gathers heat in outer space to condense and cool singularity in the centre of whatever spins. The 4 and 3 forms 7 dots.

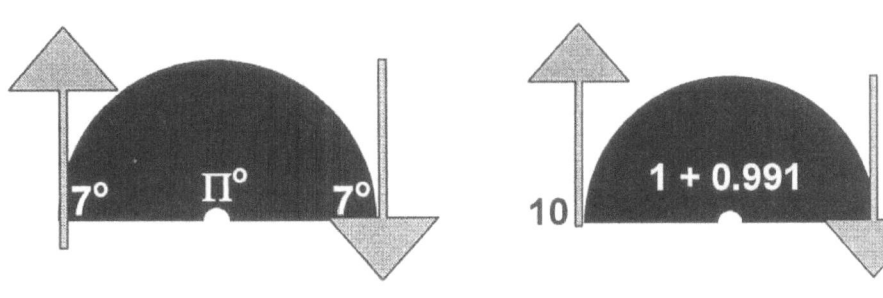

We have three points formed in the line of the axis, which I prefer to call singularity. The line only comes about when the circle turns by putting four opposing points turning in a circle around the points. In this we see that seven point three in the line of singularity and four in the circle singularity forms. That show the circle value or the value material represents in the forming of Π is 7, which we also see, is the bottom part of the equation forming Π. In the equated version of Π there is $\dfrac{21.991}{7} = \Pi$.

We do find 7° on one side of the divide turning opposite to 7° on the other side of the divide but since it is part of one circle moving both still are open and the same being the bottom part of the equation representing Π as in $\dfrac{}{7} = \Pi$. However, in the space factor the 10 on the one side is not equal to the 10 on the other side and the influence of the turning circle

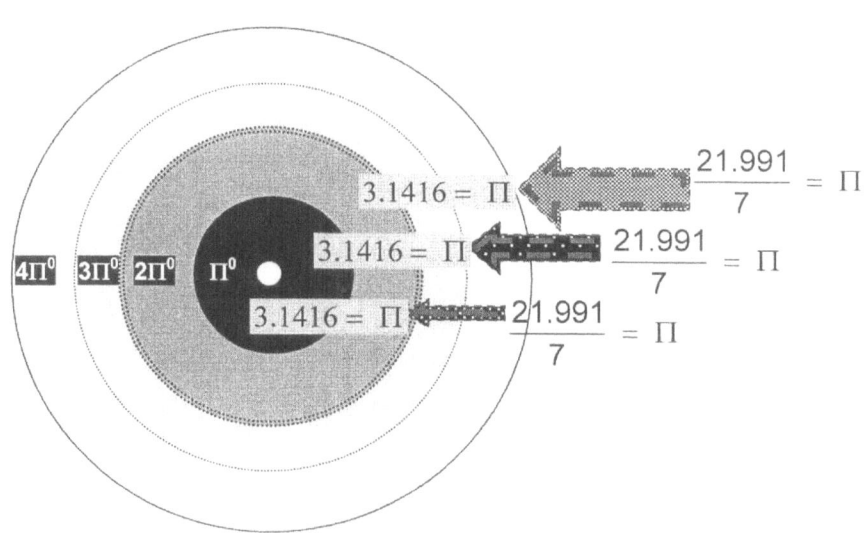
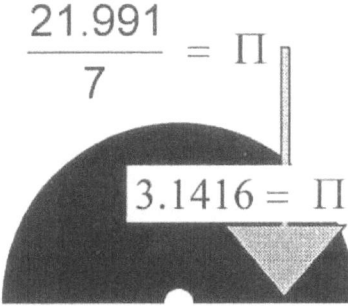

and therefore the total of space influenced by the effect of 7° on the circle is 10 + 10 = 20. By spinning the point holding singularity expands from 0.991 to become 1 on the other side if the divide. We know this because where 7 moves from the past (7) to the present (7) and onto the future (7) we have $\Pi = (7+7+7=21)$ leaving a point of growth where the Universe connects to singularity in time by 0.991. That puts on the space side a value of $\dfrac{21.991}{} = \Pi$ while the circle side is $\dfrac{}{7} = \Pi$ and the total is $\dfrac{21.991}{7} = \Pi$. Now comes the gravity part where the Coanda effect proves what gravity is.

When seven divides by seven 7 ÷ 7 as the rerouting of the circle takes place, this pulls the entire structure from space to the circle and from the circle into singularity making the entire combination go singular and the axis or singularity positions in a new time allocated place in relation to the centre that cannot move. The 21.991 becomes $\Pi = 3.1416$ while 7 becomes $\Pi^0 = 1$. Gravity forms when the cosmos goes "flat" by relevancy $\Pi^0 = 1$ and not by space and every point serving singularity goes singular.

The Earth spins and this spin influences object falling because of space compressing the atmosphere thickens. In **The Absolute Relevancy of Singularity in terms of The Sound Barrier** I elaborate on explaining more, but an object falls at a speed of, which is written according to the cosmic code applying. The seven is the curvature of the Earth, the three is time coming from the past to the present and onto the future and Π^2 is the movement of gravity going downward. If you wish to prevent falling vertically, the body has to travel horizontally faster than the decline movement is. As the space goes up, the travelling speed increases by up to $4\Pi^0$ and before it could reach $5\Pi^0$ it breaks through the Lagrangian 5 points limit on movement it breaks the Roche limit of $\Pi^2/2$, which incidentally is the sound barrier. The next limit to break is the Roche limit at $\Pi^2/4$. I explain more about this in **The Absolute Relevancy of Singularity in terms of The Sound Barrier.**

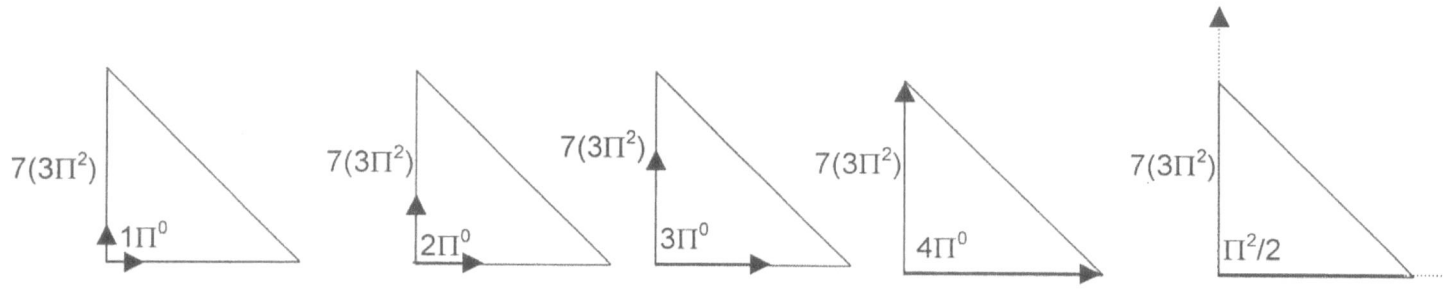

However, this is just being more informative than the level of proof this article calls for. In **The Absolute Relevancy of Singularity in terms of The Sound Barrier** I give much more conclusive explaining.

That is why we have an atmosphere. The atmosphere does not have one sole purpose and that is to carry lots of impressive Newtonian names. It is the way the earth by spin contract the space into denser Units.

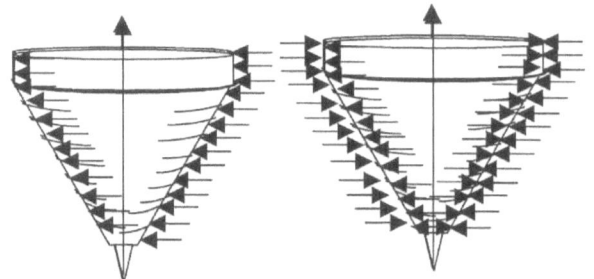

Gravity is about space contracting and not specifically air because at a height of $7(3\Pi^2)(\Pi^2/2)(\Pi^2/4)$ air has gone a long time absent. It refers to space in which the object is. However, at $7(3\Pi^2)(\Pi^2/2)(\Pi^2/4)$ the required movement to sustain flight is equal to what it is at $7(3\Pi^2)(1\Pi^0)$ because this value totally depend on the distance of movement vertically up in relation to the diameter of the Earth. The ratio of distance and density within the space compresses as the space declines from $5\Pi^0$ to $1\Pi^0$ and at $7(3\Pi^2)(\Pi^2/2)(\Pi^2/4)$ x the earth diameter there is no possibility in heaven or hell to break the sound barrier. Yet Newtonians as smart as they are, have jets fly at 31 000 meters at two-and-a-half-timers the speed of sound. This proves they are desperately short on knowledge in the gravity department because at $7(3\Pi^2)(\Pi^2/2)(\Pi^2/4)$ x Π km / h there is no sound to form a barrier which can be broken. This whole total ratio refers not to speed but to space density ranging from $1\Pi^0$ to $5\Pi^0$

I only mention this as to prove that the influence of Π^0 extends to Π as it then extends to $\Pi^0\Pi\Pi^2 = \Pi^3$.

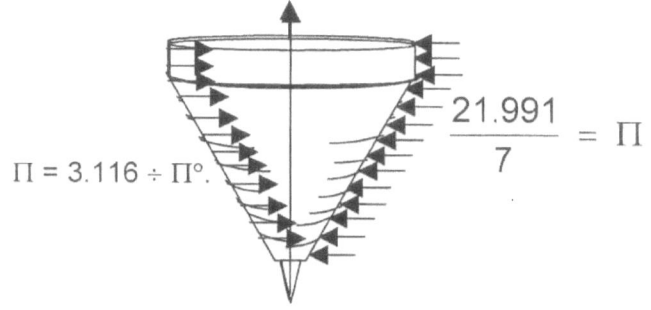

No object falls directly straight because no object moves directly straight. Everything within the entirety of the Universe moves straight by going in circles and if anything then that is what Kepler's formula $a^3 = T^2k$ indicates. Movement always goes according to Kepler's formula, which is $\Pi^3 = \Pi^2\Pi$ when in association with gravity or Π. However, movement is formed by a circle spinning while also being displaced by time in a linear direction and a sphere is a multitude of circles entwined by singularity $\Pi^0\Pi\Pi^2 = \Pi^3$ that places movement at a ratio of $\Pi^2 = \Pi^3 / \Pi$. This produces movement in space $\Pi^{-1} = \Pi^2 \div \Pi^3$ to the measure of $\frac{21.991}{7} = \Pi$ and movement of material to the measure of $\Pi = \Pi^3 \div \Pi^2$ that then becomes Π to the value of $3.1416 \div \Pi^0 = \Pi$. It is an alternating between $\frac{21.991}{7} = \Pi$ and $3.1416 \div \Pi^0 = \Pi$ interrupted by $\Pi^0 \Pi \Pi^2 = \Pi^3$.

I shall try and explain what this concept holds in terms of a piston moving while working inside an internal combustion engine. The piston goes up to a point we call top dead centre where the piston stops and according to the crank the piston halts in directional movement. Then the piston starts to accelerate to a point we call bottom dead centre where, again it comes to a dead halt. The piston stops directional movement at T.D.C. and at B.D.C. or that is what we see without seeing anything. This is not the case because if this was the case the engine must vibrate at those two points of stopping. We reason that the piston stops twice and starts moving on the two occasions (at the very top and bottom) but if that was the case of stopping at two points without stopping anywhere else, the vibration that the stopping will cause will have the engine disintegrating completely.

To us favouring positions the piston stops at two locations but the fact of the matter is that the crankshaft stops every 7° of rotation and if the crankshaft stops, then so does the piston stop. The stopping is a continuous and is an ongoing process that happen every 7° of rotation. The crankshaft moves in a straight-ahead position going straight and then it stops and redirects by 7° and then it turn by going straight again. It is $a^3 = T^2k$ and then it stops (a^3), it turns (T^2) and then again goes straight again (k) while holding reference with singularity $k^0 = a^3 \div T^2k$ all the time.

One cannot part the redirecting and the going straight T^2k because it is the same movement since the space forming a^3 is equal = to the turning T^2 and the going straight k. This is evident when dissecting Kepler's formula $a^3 = T^2k$ that $T^2 = a^3 \div k$ and $k = a^3 \div T^2$ while honouring Newton's 3rd law $k^{-1} = T^2 \div a^3$. Please believe me that this puts movement in such a perspective that it must be the most complicated dimension because this has the material $a^3 = T^2k$ moving $T^2 = a^3 \div k$ in terms of ($k = a^3 \div T^2$ as well as forming $k^{-1} = T^2 \div a^3$) while always referring to singularity $k^0 = a^3 \div T^2k$.

Kepler gave his formula symbols $a^3 = T^2k$ that do not quite represent gravity in its true symbolic nature and that then was the reason why I came on the idea that gravity has to link to Π more than any other value or symbol. It is because everything holding gravity or representing gravity (not mass) is round. Gravity connects by the use of Π. We have to part what mass does and what gravity does.

Mass is where the object connects to one point on Earth and being at that point with mass the Earth does the moving by spinning. The spinning of the Earth then represents the movement or the intention to move because the Earth spins by Π. This movement gives mass its qualities because mass does not possess the influential value of Π since mass is a quantity representative of the amount of atoms and not the spin of the atoms within the mass quantity.

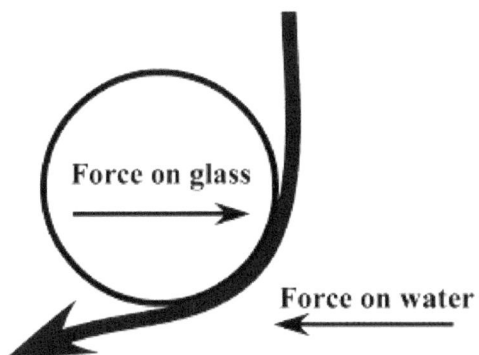

If we look at the way the Moon connects to the Earth, committing movement in a circle does it. That represents Π. When we look at the way the solar system connects to the Sun in circles every planet holds an individual symbolic value to Π that circles in relation to the Sun. If we look at the roundness of galactica, the formation represents Π. Every cosmic star holds roundness and roundness only represents one value, which is Π. The connection gravity has is not by mass but it is by Π. When we go in search of a cosmic resolve to find gravity, we better start looking for the influence Π has on the subject or leave the entire subject alone because the gateway in understanding gravity goes by the meaning of Π relating to Π^0.

The condition for the presence of this singularity that forms everything, controls everything and is everything is the centralised $k^0 = a^3 / (T^2 k)$ singularity that forms by movement $T^2 = a^3 / k$ of the space $a^3 = k T^2$ in relevancy $k = a^3 / T^2$ going both ways $k^{-1} = T^2 / a^3$ thereof (Newton's 3rd law). Now put this formula in terms of gravity and we can see the gravitational picture of the Coanda effect come to life. This also says the condition for the presence of this singularity that forms everything, controls everything and is everything is the centralised $k^0 = a^3 / (T^2 k)$ singularity that forms by movement $21.991 \div 7°$ of the space $a^3 = k T^2$ in relevancy $3.1416 \div 1$ going both ways $21.991 \div 7°$ thereof (Newton's 3rd law). Now put this formula in terms of gravity and we can see the gravitational picture of the Coanda effect come to life.

The condition for the presence of this singularity that forms everything, controls everything and is everything is the centralised $\Pi^0 = \Pi^3 / (\Pi^2 \, \Pi)$ singularity that forms by movement $\Pi^2 = \Pi^3 / \Pi$ of the space $\Pi^3 = \Pi\Pi^2$ in relevancy $\Pi = \Pi^3 / \Pi^2$ going both ways $\Pi^{-1} = \Pi^2 / \Pi^3$ thereof (Newton's 3rd law).

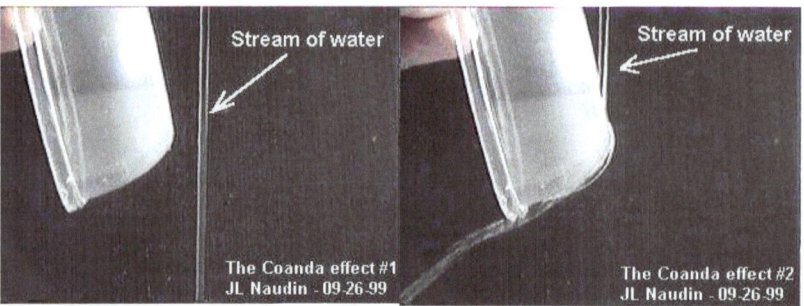

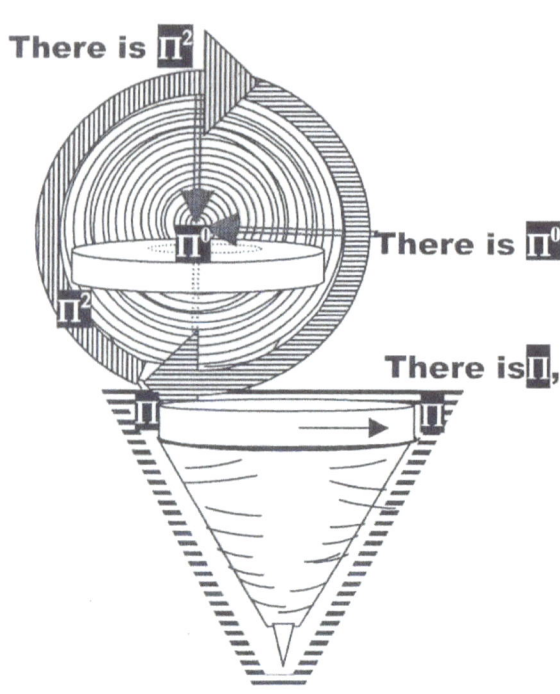

This explains the Coanda effect and **the Coanda effect is gravity** and gravity "glues" the water to the glass! The water forms a value of $\Pi^{-1} = \Pi^2 / \Pi^3$ while the glass forms a value of $\Pi = \Pi^3 / \Pi^2$ This process happens to all spinning things and as much as it happens to a piston connected to a crankshaft, just as much this will happen to an atom spinning an electron in a similar manner as the crankshaft is spinning holding a piston connected. The Universe holds infinity that parted from eternity when time began to form space. Time parted from space by parting heat from cold when eternity parted from infinity, when Π^0 singularity parted from Π singularity, when 1^0 parted from 1^1. The one point in infinity contracts while the point in eternity expands and a universe came in between. This is the foundation of gravity and this forms the foundation that built the Universe.

By contracting heat, which is what forms outer space and what all space throughout the entire Universe constitutes of, forms space that is heat that is expanded to its full. In between infinity and eternity a circle became space by the value of Π.

As much as there is expanding coming into the Universe of the atom by the motion allowing …just as much is there expanding coming into the Universe outside the atom by the motion of the atom allowing decrease in density in outer space in material…. Which puts growth in material…. That brings growth in space by density reduction in outer space …and the growth in heat within material is caused by the reduction of density of heat in outer space and in this the one loss compensates for the other gaining. Material controlled by movement (the atom) and material left uncontrolled and thereby controlling heat by expanding was precisely evenly distributed before half of the Universe started to expand and the other half stared to contract the half that was expanding. The evidence of this we find in the forming of the Roche limit and the value we derive from the Roche limit. In the end the two factors that formed will again be evenly distributed when the unification of eternity with infinity places everything back into singularity as is happening in the Black Hole at present

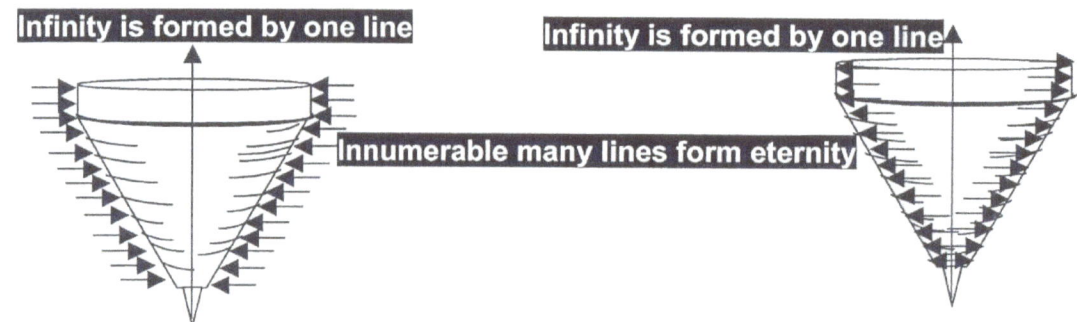

The atom becomes the dead giveaway in supplying evidence supporting the idea that spin gradually relented as space grew to become more accommodating. Time is the relation that eternity forms with infinity. Physics to this day never tried to establish where infinity in the cosmos is and physics ignore the location of eternity on a daily basis…and then Newtonians feel surprised that they don't know what time is!

The Coanda effect exemplifies the definition of time or gravity, which is the same thing. Time is the movement of everything in relation to one specific point or gravity is the movement of eternity in relation to infinity. In the centre of the spinning top there runs a line that connect space on the one side of space to the other side without having any space within that centre at all. On the outside we have space in the form of heat which is cosmic gas and by spin it condenses to cosmic liquid, which the spinning of the top turns from being a gas to being a liquid at the very edge of the material rim. That liquid is eternity because that liquid never ends being present throughout the entire Universe. The form may change from being a gas as it is in outer space to being a liquid as it is in photon light and being a solid as we find in atomic construction but it is the very same thing forming the entire Universe it is singularity. By the relative movement the singularity forms the Universe into whatever density the material may hold, but eventually it still forms by one substance and that is singularity.

In order to understand time we have to return to the top because it is through the ration of the top that we can learn so much more about the characteristics we find as gravity as it is practised by the atom. To give the top cosmic viability one must throw the top in and that motion supplied to the top, the throw initiates a time line within the top. After all it is gravity that keeps the top as it is spinning in an upright position while it is spinning because it is gravity that stabilises the cosmos. Moreover, what is actually in progress from the top spinning is the Coanda principle activating gravity and that happens in accordance with Kepler's formula.

Where we go now is the very point where space forms. Do not think of anything visible because it is where singularity announces space. It is coming into space at the very rim of space. Do not think of an atom because where we now go, the atom is as large as the largest galactica compared to singularity meeting space at the point that we venture. Looking at the way time forms we find that every point on the circle rotating from such a point it holds, when it leaves that point going to the next point and this includes every other point there may be in the entire Universe, the new point will be opposing any other point from where it came that was not pointing in the direction to which the first point is pointing, whereby it diverts from the direction it should extend and changes by 7° the direction it holds. No matter what the point is or where the point leads, such a point holding a specific direction will be unique in the direction it is rotating because at that or any other specific point wherever, it will be directing not in the direction it travels but spins in the direction flowing from the centre point outwards.

Any point will be opposing itself within the rotating of 180° changes every aspect of its previous flowing characteristics it previously had or will once again have in 180° from there. This makes everything in the Universe go in circles or become cyclic or seasonal. This is why planets don't bash into the sun like they should have if Newton's mythical formula $F = G \frac{M_1 M_2}{r^2}$ did apply. Everything repeats what was taking it into the future again and again While in rotation from the point of an outside observer all may seem static and never changing but to the object in spin every next second will be a diverting from every aspect it was in every second passing, and the direction it held in relation to the direction it held the previous mille, mille second will totally be incompatible with the direction it holds the very next mille, mille second of rotation. That proves no point can be static or constant, all though it may seem that way to outsiders. In every repeat of what was to again come in place, a subtle change comes about because of cosmic growth taking place.

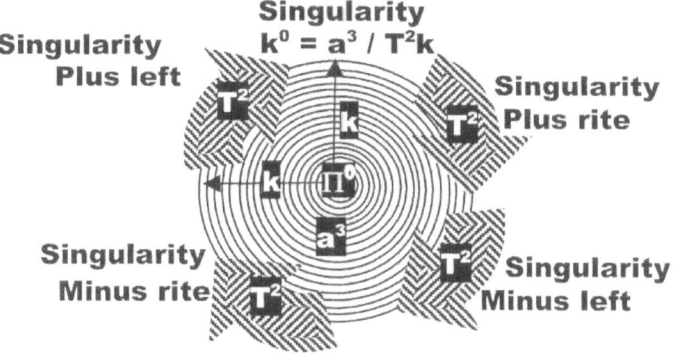

To have singularity is to have gravity but to have gravity there has to be a point of motion and a point of sturdiness. The point of sturdy is in the centre of singularity, but then the solid must be motion. However, what moves forms liquid in the presence of a solid and at that point singularity presented the solid therefore what we might think of as solid was the liquid because it moved around the solid. Where the one factor is duplicating the other factor is **compressing $k=a^3/T^2$ as well as $k^{-1}=T^2/a^3$**. Where liquid has a ratio to Π of $\frac{21.991}{7} = \Pi$, solid has the ratio to Π of 3.1416 ÷ Π^0. As one can see with the spinning top is applying the very same principle we detect what forms the earth atmosphere, which is delivering the Coanda principle. It is overheating that brings about expanding or movement and movement brings about cooling and in that the Universe started to be what it became and still is today where every point by not moving will be overheating which brings about movement that can spawn space-time by centralising singularity.

The motion is purely is space-time duplicating and the duplicating is feeding heat to the centre from the four points overheating thus the points that shows expanding. But also the duplication leads to the spawning of one point of singularity that provides the installing of the next centre for the next sphere that will result as time duplicates the sphere to a new location. Because of the principal in which the Coanda works the motion will centralise a new sphere and by appointing six positions around the centre therefore three points will not move while four will move about the three points forming the centre line. The result is that the four points by duplication will reserve the point moving as the next point in singularity because of **$k=a^3/T^2$** singularity will be a natural result of the motion. Then that point will secure a position **$k^{-1}=T^2/a^3$** which will secure six points about such a centre. The centre will bring about four points spinning around three points holding a line forming singularity. The line in singularity will stand in relevance to the contacting factor and the duplicating by expanding points will be four and serve the relevancy by contributing **$k=a^3/T^2$** as space-time only in form.

This proves that gravity is the Coanda effect and in another book I prove that the Coanda effect has its origins in Π forming a value and that value forms gravity. In order to understand physics applying in cosmology I had to start by dissecting the set-up forming pi. Using this argument I can introduce my theory on the **Absolute Relevancy of Singularity.** At the point in the centre of the circle a line must start. In the beginning when I explained the way I figured how the line starts I said a lot of dots has to continue in order to form a line. It would be 1 + 1 + 1 etc. because the line must form by holding singularity. After that point does mathematics begin but in the line that forms representing space as all other factors, then time holds 1. The line can only form when all the points forming the line have the value of 1 being 1^0. In that conclusion one realises something must separate singularity from all other factors because singularity hosts all other factors but is by own initiative Π^0.

Only when singularity meets the end value can the end value have Π where the final ring of the spinning circle forms $\Pi\Pi^2$. That will be the spot of origin forming the relevance in Π. That will hold the eternal spot…the smallest spot ever because all spots that ever can be were secured in a position in the centre of that spot that must continue as a line that forms. Because of the progress singularity follows from the single dimension singularity only allows mathematics a start at Π^0 progressing further onto Πr^0 and from there the line is born as $\Pi^0\Pi^0\Pi^0$ and to $\Pi^0\Pi^0\Pi^0\Pi^0$ etc. where Π^0 then may form the concept and value of r. But the line starts at $\Pi^0 = r^0$. This forms because cosmology is singularity based and the value is $\Pi\Pi^0$. This line $\Pi^0\Pi^0\Pi^0$ of singularity can only continue because every spinning atom preserves Π^0 in the very centre and since $\Pi^0 = \Pi^0 = \Pi^0$ and is represented in the circle of every atom spinning, the line is the same without finding conclusion except at the end where it forms mass at Π. At the point where Π forms, the movement Π^2 of the circle defines the space Π^3 of the circle and it confirms the centre Π^0 of the circle through the rotation going through the atoms. Let's call this the solid forming or if you wish, let's call it Kepler's singularity. After that singularity forms a line $\Pi^0 = \Pi^0 = \Pi^0$ where this forms another line

I prove that gravity is the result of four cosmic phenomena interacting to form the value of Π which by movement becomes the value of gravity Π^2 and gravity is equal to cosmic time applying. In order to understand the development of the cosmos and moreover the start of the cosmos and the progress in the cosmos as the cosmos formed, one has to understand the measure of Π. One has to see that Π is not merely 22 over 7 or that Π is a ratio that no one ever bothered to clarify, but Π is the key that unlocks every lock that hides a secret in origins of the Universe. One has to microscopically dissect the measure of Π to find the cosmos in measure. One has to understand where 7 fits in Π. The fact that Π is 7 at the bottom and that 7 relates to a double value of 10 is a key issue.

Furthermore, it is very important to see why Π is 10 times two by adding 1.991 on the top part of the equation. In this measured value is what holds the building blocks of the entirety we call the Universe. It is behind Π that we will find the four phenomena, which I named the four pillars performing as gravity as they form gravity. It is by the actions of Π that the Universe develops. The Hubble expanding goes by implementing gravity as Π in the square through the four pillars on which gravity and time rests. It is behind Π we discover the meaning of singularity and how singularity forms the absolute and only building block as a form that forms the Universe. It is in Π we find the Cosmic Code unlocking the meaning of the Universe. Time is centralised in Π^0 that forms Π as space's limit that becomes space by gravity being Π^2.

Space is time gone to the past in which time confirms its presence it had in the cosmos by moving from the present time into space and then onto the future leaving space behind as the past. The proof of this is again the top standing upright. Time places the governing singularity Π^o in the centre and then the very next instant the governing singularity of the previous instant moves outward to form Πr^o which then keeps on moving outward as time goes on the finally establish $\Pi\Pi^2 = \Pi^3$. That is how the top keeps erect and that is how the Coanda principle "glues" the liquid to the round solid surface by guiding the liquid as it flows in relation to the round solid. The spinning top is the manifestation of the Coanda effect, which is the coming together of the Roche limit, the Titius Bode law and the Lagrangian points.

By forming a present Π^o, time is in infinity forming singularity that then has to move on Πr^o and in doing so it leaves a legacy behind being space which will form as $\Pi = \Pi^3 \div \Pi^2$. Time is the movement of everything forming the Universe where in time the movement of time relocates everything in space by moving from the present onto the past leaving behind space as a history of time gone by. That way the top can stay erect. As time becomes the past by going to the future it forms space as it confirms the past, and in that space is what time forms by going to the past leaving space behind. Space becomes what time was at the point where time formed the particular space in relation to Π. As time becomes the present coming from the past, time has to move on to the future by replacing the past with the present at the same time and as time moved on it left space that represents that instant in time in relation to other space that was in some position at a specific location at such a point in time wherever that point in relevancy might be. The fact of Π not only refers to form but also validates the Universe by splitting infinity from eternity. By forming space when creating Π, time is using Π^0 in establishing movement Π^2. It is in the process of relocating Π to new positions by establishing Π^2 and connecting this as it forms a network consisting of Π^o by forming space Π^3 in relation Π that establishes infinity Π^o that always stays motionless. If not for movement, the Universe would be one line holding time by repeating singularity Π^o uninterrupted and it is in the diverting of eternity to a position away from infinity that the Universe comes about. This is what happens in a Black Hole where no movement within the Black Hole places eternity that always moves in a standing position to infinity that never moves. Without movement the entire Universe will fall back into and onto one point and everything we thought is real and solid will disappear into that one point holding infinity onto eternity where infinity and eternity then reunites without holding space by any measure thereof. This proves the Universe to be an unreal concept with space being no reality at all but for the movement of space in relation to singularity Π^o whereby Π confirms everything in a location in relevancy to all other things in a specific time slot or space.

When I, as a person forms a part of the Earth by the virtue of having mass that connects me Π to the Earth Π^2, stands on the Earth Π^3, my position in relation to the Earth gives me a specific positional relation to time Π^0 and the Earth. That gives the Moon a future of say one point five seconds being the past in relation to the Earth and that gives the Earth a past in reference to the Moon's future of one point five seconds. Where I am at any specific point in the present, that point I am holding is that which secures my present point in time. The Sun is eight and a half minutes into my past with all the space being in-between the Earth and the Sun and by my view of the Sun I have a present time slot, as it also gives me a past of eight and a half minutes in relation to the Sun since the light travelled eight and a half minutes through space to confirm my past during that present instant.

That secures my past by eight and a half minutes at the point of giving me a present location in time. However, that also secures my future I have from the point I now have in the present by the margin of eight and a half minutes because that establishes a flow of light that would last another eight and a half minutes of filling a presence worth eight and a half minutes while travelling through space by moving with time and every spot filled on the way would secure a position that I will have in a future presence for the next eight and a half minutes, which then becomes my future as it fills my past.

Looking at this scenario in a view from Alfa Centauri the allocated position Alfa Centauri holds in space relating to the Earth, gives the Earth a past of say four point six years while this secures the present and having that present secure the Earth to a future of say four point six years by forming time as space between Alfa Centauri and the Earth and this is confirming time to the tune of four point six years. By securing movement it forms time in having a past in relation to the present that by the same margin also secures a future in relation to a definite past. This is how the Universe builds space in establishing time. This applies to all allocated positions of rotating objects throughout the Universe. This means that every point away from Π^o serving as Π, wherever that might be, secures the past the cosmos and I have by giving the cosmos and me a future in terms of the present Π^o.

Take this in relation to Kepler's formula we then find the Earth (a^3), which is in relation as viewed from Alfa Centauri (**k**) four point six years (T^2). That secures the three dimensional status the Earth has ($a^3 = T^2\ k$) within the space from the Earth to Alfa Centauri (a^3) forming the Universe in terms of a present (k^0) being in the Earth centre which then depends on a location (**k**) secured by a future (T^2) that will come by movement where the future also doubles as a past (**k** $= a^3 + T^2$ and $k^{-1} = T^2 + a^3$). That is time and that is how time forms space and that is how space-time forms the Universe and that is the ***Absolute Relevancy of Singularity***. That then forms time in the centre in infinity in relation to space in eternity in singularity where time that moves forms space by holding time that does not move secured in positions in relevance to where every point that previously formed was in space which is time that has gone by. $\Pi°\Pi$ **divides** **infinity**$\Pi°$ from **eternity**Π where **infinity**$\Pi°$ can't **move**Π^2 and **eternity**Π eternally moves as time$\Pi°\Pi$ that establishes space Π^3 in motion$\Pi\Pi^2$.

If we put this in terms of singularity (Π^0) we find the Earth (Π^3) is in relation as viewed from Alfa Centauri (Π) four point six years (Π^2) while moving in that space that is time that has gone by. That secures the three dimensional status the Earth has (Π^3) in terms of a present (Π^0) that depends on a location (Π) secured by a future (Π^2) that will come by movement where the future ($\Pi = \Pi^3 + \Pi^2$) moving forward that also doubles as a past ($\Pi^{-1} = \Pi^2 + \Pi^3$) by the light coming from and thereby confirming the past. That is space formed three dimensionally by keeping time in infinity apart from time in eternity. The relevance (Π) that forms in relation to the present (Π^0) will relate to movement (Π^2) and the movement is circular which ensures that the relevancy forming is circular (Π) by securing that the movement is circular (Π^2) in terms of one specific point (Π^0) in infinity which then secures a roundness (Π^3) that forms an everlasting eternity ($\Pi\Pi^2$) which validates a never ending circleΠ^3. In this time in infinity (Π^0) that secures that there is an everlasting eternity ($\Pi\Pi^2$) in space (Π^3), it is not the space that is everlasting but the movement of time by the line ($\Pi\Pi^2$) that is everlasting.

Time is the movement of space in relation to any one centralised point not spinning securing such movement. Everything in the Universe moves in relation to any one single point and every one single point that forms in any location everywhere that then has to stand still to form the centre of the Universe wherefrom that point must be motionlessness to allow everything else movement. The point not moving is anywhere and the rest that moves is everything excluding that one specific point that is motionless. In that manner the Universe is constructed and with every point being confirmed only by the movement of all other points around any specific point that means there is no valid solid Universe because the Universe is constructed from singularity (Π^0) that holds no valid space (Π^3) other than being in position (Π) at a specific point ($\Pi°\Pi$) while having gravity (Π^2) that forms the time (Π^2), which is also the movement (Π^2) which is gravity of space (Π^3).

In this I prove that for instance amongst so many other things that electricity and gravity is the same thing. By ticking $\Pi°\Pi$ time forms space by becoming space as time moves into the future leaving the past behind as space. Time is a substance and the only renewable substance with the ability to come into the Universe because from the start it came into the Universe to form the Universe as space. As time moves on space grows by the margin of singularity $\Pi°\Pi$ leaving spots that form dots. The proof of this is in the value of Π being 3.14159 where 3.14159 -3 = 0.14159 x 7 = 0.9911, which is singularity as the spot (0.9911) becoming singularity 1 as the dot. In other work I explain this in much better detail.

There are two definitions we can use when looking at such a growth. We can look at the space not holding material that grew in size in which the stars finally froze their development to end as Black Holes and the growth was in terms of reducing space by remaining behind in terms of the expanding Universe all because of a lesser developing singularity within material compacting singularity $\Pi°\Pi$. Or we can focus on the stars growing $\Pi^1 = \Pi^3 \div \Pi^2$ and with that push the outer space much more into expanding by reducing the density of outer space $\Pi^{-1} = \Pi^2 \div \Pi^3$. As the cosmos grows in space, the cosmos in expanding progresses just as much as the star was reducing in space and the space in the star that became less is the same space as that with which the cosmos expands. This ratio is the ultimate relevancy. This comes from the manner that the star manages to destroy space or dismisses space or compacts space and redirects the space to go from a gas and become more compact and denser by forming a fluid where the fluid is light or heat or the solidity of frozen space as matter really is. In the Black Hole it reduces much further as it claims the singularity, which the object had, and destroys all space and all time there ever was. As the star condenses space on the inside making the star to appear as if it is shrinking away, the space in outer space seemingly becomes more as it seems to be expanding but in real terms this is just a relevancy of one becoming denser and the other losing density. That stars get hotter towards the centre is not the pushing of mass, but is about space condensing.

The entire truth about the cosmos is that the Universe is within the atom that forms a cosmic unit holding singularity as much as it secures singularity and every atom forms a Universe standing apart, parted by time from all other atoms by the spin produced. Every Universe formed by every atom starts in infinity and ends where each atom's spin is forming relevancy between where that Universe starts and ends. All atoms are a Universe formed within the space that time puts between infinity and eternity. All atoms are stitched together by an invisible, unseen singularity - string that is present while also being absent and this invisible string links everything that the Universe is throughout the entirety. The entirety rests on relevancy. As time moves on forming a line by implementing more dots in relation to the dots that are already there forming the history of time, which is what we call space, the area we call outer space receives many dots that time leaves as a footprint while the dots time leaves within material are less, just because the space is concentrated and thereby is less. The dot that time leaves holds no space but in terms of space moving with time increasing the adding of space-less dots brings about more space which then reduces the concentration of space and the more the dots are, the more the concentration reduces. This is why the top can stand erect when spinning. It is because time forms a governing singularity Π^o that then shifts in the next instant outwards to form Π as the controlling singularity in terms of the movement Π^2 that then controls the space spinning Π^3. It is time leaving Π^o that then the next moment forms Π and in the movement of gravity Π^2 the space forms Π^3.

With more dots landing in outer space since there are more space, the space density reduces as the expanding in outer space seems to be more than what is applying to material where space is at a premium, being condensed. With time duplicating to form dots in singularity, every instant that it produces spots forming dots as the present, the space that outer space gains supersedes the space that material gains and that makes material more compact or more and more dense in relation to outer space. The space gained by the space occupied by the moving of material receives fewer dots than the space forming outer space or that part which we see as outer space and the space material holds advances more in density through the loss of density in the space called outer space.

This leaves material more compact in relevancy that seems to hold less space and this is moreover because of the relative loss of density in outer space is there because of outer space gaining space by time leaving more dots. The density in outer space is thereby lost and in that the density in material is gained by the loss of the density in space in outer space being more because it is a greater recipient of time. The dot also leaves one point every time on the dot forming the governing singularity and that confirms the point holding governing singularity in terms of many dots received by the spin of the controlling singularity in terms of the gain of endless space in outer space. In that material always grows as outer space declines in density and that forms the "Hubble Constant" that is no constant. The Hubble constant is gravity expanding, which contradicts Newton's gravity contracting.

In a nutshell that is gravity. It does not even mention mass because mass has nothing to do with Π while gravity is Π in more forms than what humans are able to imagine. The cosmos grows by gravity which is Π^o = $\Pi^3 \div \Pi^2 \Pi$.

That is why the distance between the Earth and the Moon becomes more. That is why the circumference of the Earth becomes bigger. That is why there are Earthquakes and hurricanes. That is why a human grows and heals and that is why hair and nails grow. That also is why there is aging and eventual unavoidable death to material holding life. The body never stops growing, which brings about the inevitable decline of life's body structure, as time becomes more that the body endures. The ever growing of the body makes the body collapse on itself with aging. As time goes by everything on the Earth including the Earth and everything in the Universe around the Earth is gaining in space because that is what time leaves. That is why everything in fossils seems to be bigger the further back the fossil goes in the history of the Earth. Newtonians show millipedes that once roamed the Earth that were one metre wide…and Newtonians not only believe that but also advocate this information as the truth! Everything holding material grows by time leaving space as the history of time that went by. The history or the space of the millipede became bigger as time moved on but the millipede never was one metre wide. That is why we can see galactica so far way. It is through time progressing in space that it carries light to move from there to where we are capable to see where the light came from. Time brings light all the way by progressing in space that carries light through space.

There is and there can be no such a thing as "dark matter" What would make matter "dark"? If the material is "light" it then has a higher concentration of light than where we are at present. This puts the object we see in a denser area than where we are. There is much more movement in that area that concentrates the space in that area and thereby we can see the area because the space released from that area expands as it comes towards us and that light expanding is what we visually see. On the other hand areas that seem dark are more expanded with our light flowing outwards to those areas. Being darker is having light flowing to that area from where we are. That puts that object in a more expanded environment and in higher expanded surrounding than where we are.

If the material is "dark" our light is moving towards that position and that makes that area move slower than what we do. That area is therefore less concentrated and more expanded than where we are. Then again if we see the area as light, the area is more concentrated in density having light poring out towards us by the measure of releasing density. The light flowing towards us will make the region seem as if it is lighter. It again is about relevancies. The part that seems to have brighter light moves faster as the light moves at a greater pace and moves towards us. The area that seems to be darker has light moving much slower because the light is moving away from us as it is the light from our area expanding into that larger area that leaves us with the concept that that area is darker.

As the light moves into an expanded area it will seem to slow down. It is a question of relevancies applying by movement in relation to "standing still" or "moving faster" and "moving slower". If we look at the Earth from the Sun the Earth where we now are would be so dark where the earth is located that Earth would be invisible from the Sun because our space where the Earth is there is so much more expanded than the space is that surrounds the Sun. Again seen from the Earth when looking at Pluto but by only using the naked eye Pluto is so dark it is "invisible" to the normal human perception. It is because "space" is much more expanded out there than it is where we are and if it is more expanded it is moving in relation to the space being available in which it can move making movement seem slower and making that area seem bigger. Then we have Mercury of the approximate same size but are very visible because it is more compressed in that area and therefore more visible than where we are and with the larger density the reflection of the Sun seems to light up the planet.

My question coming from this is why there is this hunt to find dark matter. Dark matter there is because dark matter is only more expanded in terms of denser matter which has light flowing to us which makes us able to see the light coming to us. I am the first to admit that there is no substantiating proof presented in this article alone and I don't even begin to claim that I deliver any proof in this article. There is no room to present even the least bit of proof in any form possible in the space given to this article. With the limited space available to publish information in a journal by way of a small article such as this and having so much information at a premium I decided to release some vital information and the required proof about my claims in other small but comprehensive works that can be obtained.

By going to LULU.com the following books are available in e-book format as individual books wherein I share with you the newly discovered information. For more information visit www.LULU.com and search for www.sirnewtonsfraud.com and see how academics use Newtonian disinformation to brainwash students in physics. The download of www.sirnewtonsfraud.com because I wish to put a stop to such counter productive practises. If you are a practising physics lector or Professor I would take any challenge where you can show what I accuse you of is not totally valid. You brainwash students into accepting as much as believing

$$F = G \frac{M_1 M_2}{r^2}$$

is the truth. Read www.sirnewtonsfraud.com and get a book load of questions whereby you barrage them into getting honest about physics for the first time since Newton. Let them prove mass plays a part in gravity or admit they falsify formulas to bedazzle the minds of youths.

Visit www.LULU.com and search for:
The Absolute Relevancy of Singularity The (proposed) Article
I purposely withhold the publishing of the Absolute Relevancy of Singularity: The Article until publication of the article runs its commercial viability as it is published in the journal.
The Absolute Relevancy of Singularity The Dissertation
The Absolute Relevancy of Singularity in terms of Applying Physics
The Absolute Relevancy of Singularity in terms of The Four Cosmic Phenomena
The Absolute Relevancy of Singularity in terms of The Sound Barrier
The Absolute Relevancy of Singularity in terms of The Cosmic Code
www.singularityrelavancy.com

The Absolute Relevancy of Singularity The Article is written as the first introduction to introduce singularity that forms gravity in the new theorem explaining the Absolute Relevancy of Singularity. Since the article was comprehensive but was adjudged as to long for a physics journal, I decided to offer the article in its original and total layout in which I introduce the framework of my ideas.

The Absolute Relevancy of Singularity The Dissertation is there written as the second introduction to introduce the four pillars in a very wide sense on which the new theorem rests. This is to convince readers about the authenticity behind the explaining and the thinking that forms the new approach to physics backing the Absolute Relevancy of Singularity where gravity depends on Π.

Then The Absolute Relevancy of Singularity consists of a four individual part theses each forming a thesis. There are either six individual books on offer in e-book format or in print could only be purchased as one unit named The Absolute Relevancy of Singularity The Theses This consist of

I explain where the cosmos starts, showing mathematically at what point does gravity start and end and why the atom forms while the atom is responsible for taking the Universe into a three dimensional form, which is at that very same point where the cosmos starts. I give a detailed explaining on this issue bringing indisputable proof in presenting unwavering facts. *The Cosmic Code as the Absolute Relevancy of Singularity* shows precisely what gravity is and why gravity and electricity is the very same thing. I show the process by which the birth of stars forms within galactica and why stars will and does end their life as Black Holes.

There are six individual e-books available from Lulu.com, which I send you on the CD. If you will read it what you are going to read when you read the work introduced in this letter is new to all of mankind. Everything you are going to encounter in my work is the fruit of my mental ability and my brain bashing and therefore was never written before by any person. Whether you are, albeit the most accomplished physicist or a first year student fresh out of school, you have the same background knowledge about the work you will encounter. It is advisable therefore for any reader to first get acquainted with the first, and then the second book then the third book and so on offered in the series before taking the challenge in reading **THE COSMIC CODE.**

This is not a ploy but this warning comes on the grounds that I use certain ideas in **THE COSMIC CODE** which I explain in the other three in the series and I do not return again to the introducing of the concepts as a whole in **THE COSMIC CODE.** Then again…these books when read as a theses shows that cosmology read correctly is so simple that even a person with a simple mind such as I have can understand all information that explains the Universe because I am not superior in even the least aspect of life. What I see is out there for everyone to see. Therefore I admit that if I could write the work on offer taking into consideration my simple level of thoughts, then any body walking the Earth can read the work I offer because I have a simple mind that offers just as simple approach to science as any child would have…but please take note that what you are about to read…that information you have never read before and in that comes the complexity of understanding and not the work that is complicated. As the books introduce new facts using new arguments and new concepts, which the readers have never come across before, so the understanding about the new concepts will grow in clarity.

This was The Absolute Relevancy of Singularity The (proposed) Article
Visit **www.LULU.com** and search for:
I purposely withhold the publishing of *the* **Absolute Relevancy of Singularity: The Article until publication of the article runs its commercial viability as it is published in the journal.**
The Absolute Relevancy of Singularity The Dissertation
The Absolute Relevancy of Singularity in terms of Applying Physics
The Absolute Relevancy of Singularity in terms of The Four Cosmic Phenomena
The Absolute Relevancy of Singularity in terms of The Sound Barrier
The Absolute Relevancy of Singularity in terms of The Cosmic Code

All I want is what Tyco Brae wanted and that is not to have lived in vain. It took forty years of study that I spent to gain the insight I achieved in one lifetime of accumulating facts and those masterminds incapable of understanding one word that I explain, decides what my research should show and what my findings should be. That which they can't understand is therefore invalid as science and must be pointless. All there is also is everything they know. In their superiority, what they don't know does not exist. By the narcissism shown and the arrogance of semi blind, self-opinionated, self serving academics in physics in office today that think they have the authority to decide what science is and what science has to be and what confirms science as much as what conforms science will decide my fait by discarding my work just because they can't understand my labour. That's what they refuse to understand.

Space is the confirmation that everything that never could have started, had started when that which can never end parted from that which can never start. Only space can be able to end because then there is only an end to that which could never have started to begin with in the very beginning. Time is the confirmation that there is no end possible to what is the present because only space can end since space started when infinity parted with eternity. It is because space started when time parted infinity from eternity that now also it is space that can end, and that is because space began and grew from where it started and by losing what never could be, only that which started could ever end. But because space started as something with no worth, and is something everything can travel through by only using time, such worth being worthless can be lost and that evidence we have by looking at the Black Hole. Time can only reunify that which parted when the perfect became imperfect and that evidence we see when looking at the Black Hole. By rejoining infinity with eternity, as that is what happens the Black Hole, space ends in a singular spot that holds time in the present and that never ends because the singular spot is time. Space is the commodity with which singularity extends its duration as it forms light by which it is forming the past for all to see within the present as the present happening presently and in that is all the value space have, it extends singularity forming a presence. With this, there then can only be the limiting of space as a value in validity. Time uses the space it takes to start the ending because everything that starts claims the space it takes to find the end while time is the process where some space is used to find the start of the end that can never take place.

The Universe will end where it started but not as it started. The liquid part will create space by losing density. The solid grows in density and reduces space by creating relevancy where space in outer space creates space as the solid absorbs what the liquid deletes while also dismissing the space within by compressing density within. Time will move more rapid as space collapses in more and more Black Holes coming about. Every star shining today will be a massive Black Hole absorbing space by reuniting time. That is the function of all stars going on their way becoming Black Holes. The process will never end, but will continue as motion, until time reunifies. This will come when material growth expands by filling all space that has vacated what never actually was in place. With this filling of space serving as material, motion then joins with what can never move by connecting what has no outside and no end, with what never can have an inside or can never begin.

This that I have written in this what you still must read puts all that was written before, and all of that which you had read, that in the past. This that is written in this what you still have to read is the future and whether you, as the reader where by that you also become the witness, decide to put this in your present, is up to you to choose. The reader must choose where to stand on these issues that is written in this what you have to read, because this that is written in these books is the future. When the future will become the present is solely the choice of the reader. Where the reader chooses to not choose what this represents then the choice will condemn the person choosing, as one that will remain with the past. The choice will place the person choosing not to choose this as the future, to become part of those who will be forgotten as the past gone to darkness and forming part of those that was not worth remembering. Such a choice on the future will put the person choosing not to choose this as the present, part of the past that eventually will become the part that was not worth recollecting. The future forming the present only remembers the past when the past is worth to be used in the present as the present. The future never carries failures in the past through to the present by remembering it in the future as a worthwhile past. Recognising this what is written in this what you still have to read, the future will place the person as a part of the future by removing failures in the past from the present and placing the future in the person's future. Realising what this replaces and then replacing this with what was, is the method of making what is written in these books becomes part of the reader's present. These books are the future and no suppressing of the powerful and the mighty wise can prevent these books from filling the future. The mighty wise can push their present into forming the past by clinging onto the worthless they represent as worthwhile that will become the past when these books becomes the present. However, that choice will doom them into the past along with all other things and thoughts not worth the burden to take into the present and onto the future from the past. Their adopting the worthless and not adapting to the truth moves them to the past only worth to be forgotten as the worthless part of the past. The choice is theirs to make.

The **Titius Bode law** has been around for centuries and with all the mathematical splendour available there for all to use, all the brilliant mathematicians could never come close to show any ability to any understanding of this very important phenomena. They could mathematically equate the formula, but then after that human intellect dries up.

The **Roche limit** has been around for centuries and with all the mathematical splendour available to apply in order to fathom concepts behind this phenomenon, still with all the computing ability of a machine all those physicists with all the mathematical superiority could not touch any understanding about the concept forming the background.

The **Lagrangian points** has been known to science for centuries and with all the mathematical splendour available not one calculation could ever explain why this event is taking place.

The **Coanda effect** has powered turbine engines and aeroplanes in flight for almost a century and with all the mathematical splendour available to design the most terrific aircraft, not one engineer could mathematically compute one fact to show understanding why this takes place. How sad it is that they remain just more computing power.

That does not say much for the bountiful prestige that mathematician's claim as their lawful bragging rights in areas where true human intellect is called on. Is it not high time to begin to admit you are playing the game of fools with you arrogance about your achievements using mathematics when designing space whirls and travelling to galactica while not even understanding what movement asks for. You do not even understand the neutron and the neutron is compressing density increasing, which is what gravity is, which is what time is, which is what all movement is…that is why the neutron has no mass because mass is the principle coming about where independent movement ends.

You're mathematics could not get you any closer than playing games in a fairy tale Universe using misguided presumptions about mass forming gravity and living the Universal farce which Newton created because that fairy land is what all the Kings clever heroes and all the King's splendid wise could never prove in hundreds of years.

If you feel superior as a scientist practising physics on the highest level having a gloating hail of superior mental capability covering you like an aura, then I have very saddening news for you. If you have the ability to compute and calculate at the highest level, then look at your computer and see one that machine has abilities as a machine which is equal to you, but it's a manmade machine. Stop playing games by creating fairy worlds making up fairy tales about fairies and little people, mass that can create forces, four of them no less, and come and join the rest of us living in reality that does not need to compute forces to be able to not understand what it is that you compute, but to use human intelligence and in that way to understand what only human intellect could ever understand.

If dear Prof Friedrich W. Hehl and all the other more than one thousand five hundred academics in physics I contacted the past ten years were unable and may I add, therefore dubiously unwilling to read this letter I wrote in an article or any of the other multitude of articles I sent in letter form which I sent to all the other academics as letters that I regard to be so simple children could understand it, then how the hell will they get around to understand real tough work such as I introduce in "**Matter's Space in Time: The Thesis**" where the explaining really gets beyond any simple common understanding, such as this work display? I could have had so much more work done by now if I did not have to fight their arrogance every inch of the way and endure their restrictive narcissism. They would never admit they have not the foggiest what I said…and so they refused to read…Stop being so utterly intellectually superior while looking at the Universe from such a mentally dizzy height and start being part of the Universe. Stop gloating in all the arrogance you mathematically can muster. There is a wonderful Universe in singularity awaiting you to discover, so join us normal people that can think normally with using reason! This is a small part of what I call the cosmic calendar. Reading from the formulas the factors indicate how the cosmos developed and will still develop until the final era arrives. Again I say it is just a small part but to show all the formula and factors; that would be pointless and serve no purpose. Explaining anything of the relevancies or any part of it at this point will prove a futile venture, but I destroyed my health and heart to formulate this. This is a part of what I named the **COSMIC CALENDER**. Formulating what you see lead to a massive heart operation because for nine months I slept one hour per 24 cycle trying to formulate the cosmic calendar. I worked everyday twenty-two hours a day to write **Matter's Space in Time The Theses** containing seven books each being a thesis.

After achieving the full **Matter's Space in Time: The Theses** of which the **COSMIC CALENDER** is only a very small part afterwards my health had it because I can never physically do any work again. This work in the books I called **Matter's Space in Time: The Theses** is extremely complicated and maybe it is just me because I am not very intellectual, well not as intellectual as those are that say this is not science, but to understand it takes one hell of an effort in concentration, unlike reading this very simple, straightforward article. The **COSMIC CALENDER** is but a fraction of one chapter of one of the books and to write all seven books took just about all of life's energy out of me. I am so tired and even more tired fighting all those that use blindfolds called Newtonian science.

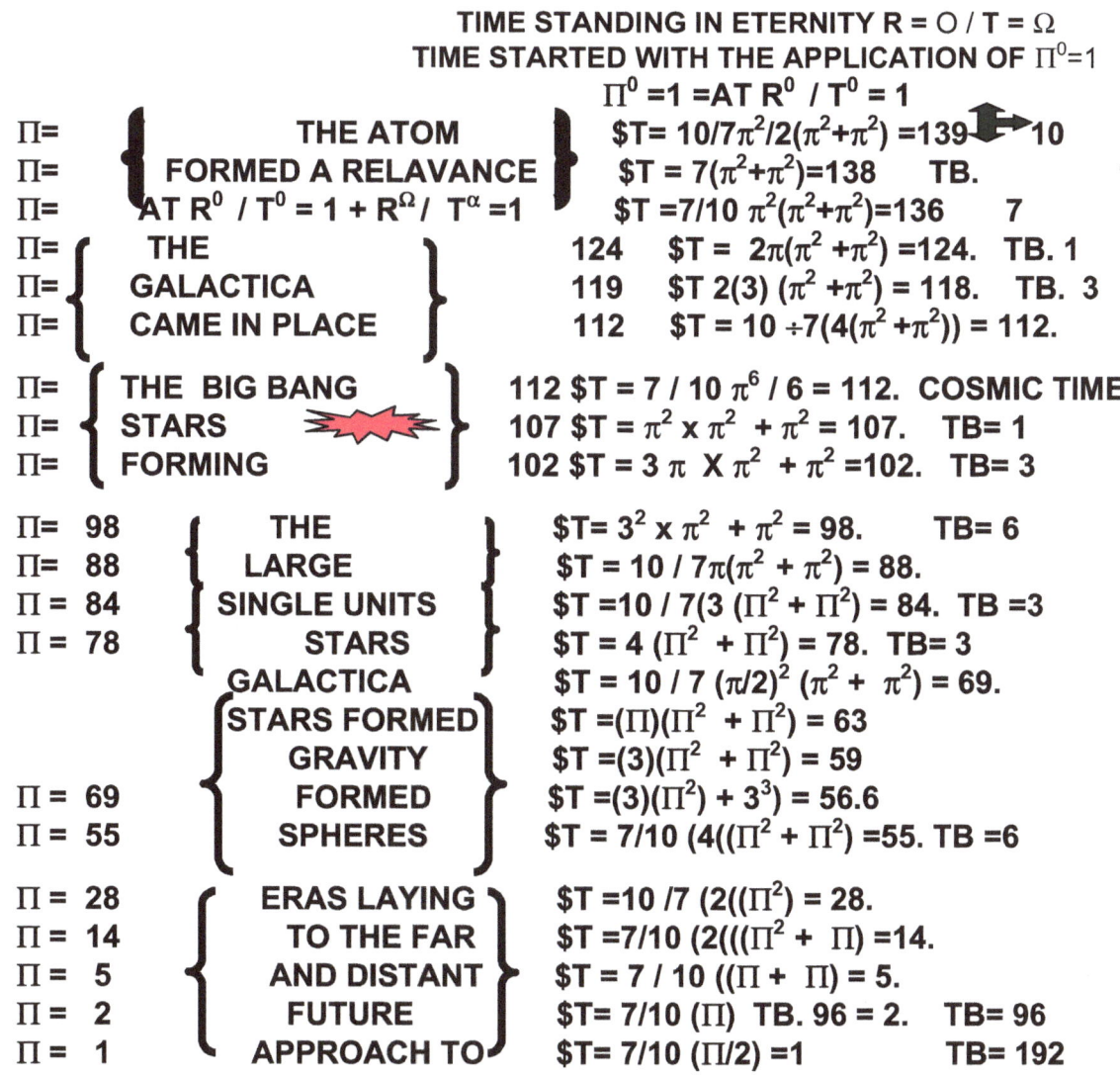

THE COSMIC CALENDER
TIME STANDING IN ETERNITY $R = O / T = \Omega$
TIME STARTED WITH THE APPLICATION OF $\Pi^0 = 1$

$\Pi^0 = 1 =$ AT $R^0 / T^0 = 1$

$\Pi=$	THE ATOM		$\$T = 10/7\pi^2/2(\pi^2+\pi^2) = 139 \rightarrow 10$	
$\Pi=$	FORMED A RELAVANCE		$\$T = 7(\pi^2+\pi^2) = 138$ TB.	1
$\Pi=$	AT $R^0 / T^0 = 1 + R^\Omega / T^\alpha = 1$		$\$T = 7/10 \, \pi^2(\pi^2+\pi^2) = 136$	7
$\Pi=$	THE	124	$\$T = 2\pi(\pi^2+\pi^2) = 124$. TB. 1	
$\Pi=$	GALACTICA	119	$\$T \, 2(3)(\pi^2+\pi^2) = 118$. TB. 3	
$\Pi=$	CAME IN PLACE	112	$\$T = 10 \div 7(4(\pi^2+\pi^2)) = 112$.	
$\Pi=$	THE BIG BANG	112	$\$T = 7/10 \, \pi^6 / 6 = 112$. COSMIC TIME	
$\Pi=$	STARS	107	$\$T = \pi^2 \times \pi^2 + \pi^2 = 107$. TB= 1	
$\Pi=$	FORMING	102	$\$T = 3\pi \times \pi^2 + \pi^2 = 102$. TB= 3	
$\Pi= 98$	THE		$\$T = 3^2 \times \pi^2 + \pi^2 = 98$. TB= 6	
$\Pi= 88$	LARGE		$\$T = 10/7\pi(\pi^2 + \pi^2) = 88$.	
$\Pi = 84$	SINGLE UNITS		$\$T = 10/7(3(\Pi^2 + \Pi^2) = 84$. TB =3	
$\Pi = 78$	STARS		$\$T = 4(\Pi^2 + \Pi^2) = 78$. TB= 3	
	GALACTICA		$\$T = 10/7 \, (\pi/2)^2 (\pi^2 + \pi^2) = 69$.	
	STARS FORMED		$\$T = (\Pi)(\Pi^2 + \Pi^2) = 63$	
	GRAVITY		$\$T = (3)(\Pi^2 + \Pi^2) = 59$	
$\Pi = 69$	FORMED		$\$T = (3)(\Pi^2) + 3^3) = 56.6$	
$\Pi = 55$	SPHERES		$\$T = 7/10 \, (4((\Pi^2 + \Pi^2) = 55$. TB =6	
$\Pi = 28$	ERAS LAYING		$\$T = 10/7 \, (2((\Pi^2) = 28$.	
$\Pi = 14$	TO THE FAR		$\$T = 7/10 \, (2(((\Pi^2 + \Pi) = 14$.	
$\Pi = 5$	AND DISTANT		$\$T = 7/10 \, ((\Pi + \Pi) = 5$.	
$\Pi = 2$	FUTURE		$\$T = 7/10 \, (\Pi)$ TB. 96 = 2. TB= 96	
$\Pi = 1$	APPROACH TO		$\$T = 7/10 \, (\Pi/2) = 1$ TB= 192	

This table shows that by removing **density** from outer space and passing on **density** to material, outer space relaxes the concentration ratio there is between outer space and material space and the relaxing comes by increasing volumetric space. By atoms spinning, the material pumps in heat into atomic space and expands material by removing heat from outer space and this will decrease heat in outer space density as it increases heat in material space and increase material density. By increasing space, material although growing in size is shrinking in space and this comes about by increasing outer space in volume size as this reducing outer space density reduces material space in ratio to the increase in the space ratio of outer space. As outer space increases in volume, so also does material space increase in volume but material space lags behind in ratio of increase and therefore material space decreases in ratio as outer space increases in ratio. Although in density material space is increasing exponentially, in the ratio applying outer space is pushing material into the oblivious and therefore making the atoms gaining in material to shrink into obscurity where all stars will eventually become Black Holes, not because they grow into it but because outer space in growth pushes the stars era by era into obscurity. By outer space losing density it is growing in worthless space. By gaining in space becoming denser, the material is losing space in ratio to the outer space gaining in space and as this ratio increases, so does the gravity within stars increase until outer space pushes all material space into Black Holes. Do the so overqualified mathematicians in physics conclude that this is the way that the Universe performs…no they have the Universe expanding? How can something that has everything such as the Universe have and within that something is everything that ever could be and in that something is everything that anything could have, then by that find more in order to expand and become bigger, meaning it becomes more. It already has everything there ever could be and now they say it is expanding by becoming more. Where is the Universe getting more to expand with and become more? This is the crap logic one gets from simplifying idiotic thoughts with mathematics. They calculate by inventing more equation in mathematics that something is getting more, $v = H_o r$, while that something being the Universe will never be able to get more since it has all it could ever have. Where serious human understanding prevails it is better to leave what belongs to machines to be calculated by machines and have human intellect bring understanding to human issues.

They fail to realise many things but mostly they fail to realise that when accepting my point of view about cosmology, then they need not fear at all because in physics and the way to **calculate physics there is not one** aspect about physics that changes one freckle or mole in conducting science. **The New Cosmic Theory** gives a new dimension in the way we are able to read the cosmos. Their buildings will not topple over and their dams will not suddenly burst but the sun will also not collapse into a Black Hole if the mass exceeds 1.6 units of solar mass as Robert Oppenheimer said when he, together with G. M. Volkoff, calculated that this would happen and subsequently was awarded a noble prize for declaring such crap. Using my Cosmic Code will alter our approach we have to the cosmos and our thinking will be entirely new because now we know what basis mathematics uses when forming the cosmos. It uses Π as one numerical unit. I never said there is no mass forming a yardstick in physics applying. I never denounced mass as a tool which to use in physics. I say mass pulls no gravity and mass depend on gravity forming while gravity is not the result of mass but eventually results in mass forming. This means in the Universe size has no place to be. Big and small are manmade features. I say mass as a cosmic factor forms no basis or find no reality. **Mass in the cosmos has no place.**

mailto:E-mail www.singularityrelevancy.com

P.S. J. SCHUTTE (PEET SCHUTTE)